IET POWER AND ENERGY SERIES 7

Series Editors: Prof. A.T. Johns
G. Ratcliff
J.R. Platts

Insulators for High Voltages

Other volumes in this series:

Volume 1	**Power circuit breaker theory and design** C.H. Flurscheim (Editor)
Volume 4	**Industrial microwave heating** A.C. Metaxas and R.J. Meredith
Volume 7	**Insulators for high voltages** J.S.T. Looms
Volume 8	**Variable frequency AC motor drive systems** D. Finney
Volume 10	**SF6 switchgear** H.M. Ryan and G.R. Jones
Volume 11	**Conduction and induction heating** E.J. Davies
Volume 13	**Statistical techniques for high voltage engineering** W. Hauschild and W. Mosch
Volume 14	**Uninterruptable power supplies** J. Platts and J.D. St Aubyn (Editors)
Volume 15	**Digital protection for power systems** A.T. Johns and S.K. Salman
Volume 16	**Electricity economics and planning** T.W. Berrie
Volume 18	**Vacuum switchgear** A. Greenwood
Volume 19	**Electrical safety: a guide to causes and prevention of hazards** J. Maxwell Adams
Volume 21	**Electricity distribution network design, 2nd edition** E. Lakervi and E.J. Holmes
Volume 22	**Artificial intelligence techniques in power systems** K. Warwick, A.O. Ekwue and R. Aggarwal (Editors)
Volume 24	**Power system commissioning and maintenance practice** K. Harker
Volume 25	**Engineers' handbook of industrial microwave heating** R.J. Meredith
Volume 26	**Small electric motors** H. Moczala *et al.*
Volume 27	**AC-DC power system analysis** J. Arrill and B.C. Smith
Volume 29	**High voltage direct current transmission, 2nd edition** J. Arrillaga
Volume 30	**Flexible AC Transmission Systems (FACTS)** Y-H. Song (Editor)
Volume 31	**Embedded generation** N. Jenkins *et al.*
Volume 32	**High voltage engineering and testing, 2nd edition** H.M. Ryan (Editor)
Volume 33	**Overvoltage protection of low-voltage systems, revised edition** P. Hasse
Volume 34	**The lightning flash** V. Cooray
Volume 35	**Control techniques drives and controls handbook** W. Drury (Editor)
Volume 36	**Voltage quality in electrical power systems** J. Schlabbach *et al.*
Volume 37	**Electrical steels for rotating machines** P. Beckley
Volume 38	**The electric car: development and future of battery, hybrid and fuel-cell cars** M. Westbrook
Volume 39	**Power systems electromagnetic transients simulation** J. Arrillaga and N. Watson
Volume 40	**Advances in high voltage engineering** M. Haddad and D. Warne
Volume 41	**Electrical operation of electrostatic precipitators** K. Parker
Volume 43	**Thermal power plant simulation and control** D. Flynn
Volume 44	**Economic evaluation of projects in the electricity supply industry** H. Khatib
Volume 45	**Propulsion systems for hybrid vehicles** J. Miller
Volume 46	**Distribution switchgear** S. Stewart
Volume 47	**Protection of electricity distribution networks, 2nd edition** J. Gers and E. Holmes
Volume 48	**Wood pole overhead lines** B. Wareing
Volume 49	**Electric fuses, 3rd edition** A. Wright and G. Newbery
Volume 51	**Short circuit currents** J. Schlabbach
Volume 905	**Power system protection, 4 volumes**

Insulators for High Voltages

J.S.T. Looms

The Institution of Engineering and Technology

Published by The Institution of Engineering and Technology, London, United Kingdom

First published 1988
Reprinted 1990, 2006

The Institution of Engineering and Technology
Michael Faraday House
Six Hills Way, Stevenage
Herts, SG1 2AY, United Kingdom

www.theiet.org

British Library Cataloguing in Publication Data
Looms, J.S.T.
Insulators for high voltages–
(IEE power engineering series; 7)
1. Electric insulators and insulation 2. High voltages
I. Title II. Institution of Electrical Engineers III. Series
621.319'37 TL3401

ISBN (10 digit) 0 86341 116 9
ISBN (13 digit) 978-0-86341-116-8

Printed in the UK by Short Run Press Ltd, Exeter
Reprinted in the UK by Lightning Source UK Ltd, Milton Keynes

Contents

Foreword

Although this book is written primarily with the needs of students, practising electrical engineers and researchers in mind, I hope that it will be helpful also to others who seek information about insulators. There is a mass of published work on the subject, disparate and often contradictory, which makes the search for specific answers to questions a frustrating task. The seeker should either find what he needs directly, or at least a reliable reference, in these pages.

My text is a distillation of much of a working life's accumulations. Some was acquired when I worked as a researcher for the Central Electricity Generating Board (CEGB), the UK's national electrical authority and a large user of insulators. The rest has come from my days as consultant to manufacturers concerned with the development and marketing of insulators. Both sides of the coin are presented here.

After more than a century's development insulators might be expected to have settled into final forms, but in fact the whole topic continues in a state of flux. Some of what is said here may well be proved wrong, before long. I shall be pleased, however, if I succeed in showing why insulators are made as they are and behave as they do; if my readers acquire a 'feel' for the subject; if I leave with them a flavour of the fun and excitement which have attended my own days as an insulator-man.

I have done my best to acknowledge everyone who has helped with my text and apologise for any accidental omissions. I particularly have to thank the CEGB for allowing the inclusion of some of the results obtained in their service, and the International Conference on Large Electrical Networks (CIGRE) for permitting the inclusion of many internationally acquired results of collaboration, under their aegis. IEE publications have been extensively drawn upon.

On the personal level, I express my gratitude to Peter Lambeth, my friend for many years, for a great deal of my material. I thank my 'boys' at the Central Electricity Research Laboratories (CERL) for their hard and careful work and happily recall the amicable collaboration of my colleagues in CEGB, in the national power authorities of France and Italy (EdF, ENEL), in the German, Italian and Czechoslovakian Testing Institutes (FGH, CESI, EGU) and among the manufacturers in the UK, Germany, France, Japan and the USA. We have fought technical battles but have enjoyed many good times together.

Above all I must thank my Chief, John Samuel Forrest, F.R.S., the father of insulator research. He set my hand upon this particular plough and guided me along the furrow. I dedicate this book to him.

East Molesey

J.S.T. Looms

Acknowledgments

I thank the following gentlemen for their help, in writing this book and the named bodies for their courtesy in allowing reproduction of photographs or figures:

Arthur Barber
Ewald Bauer
Alan Bradwell
Robin Clabburn
Paul Claverie
Fritz Hirsch
Ray Houlgate
Tony Howard
Alan Ingles
Gerard Leroy
Ian Maclean
Carlo Malaguti
Jack Moran
Marcel Moreau
Stan Morgan
Katsuhiko Naito
Harry New
George Orawski
Laurent Pargamin
Rene Parraud
Dick Penneck
Tony Pentelow
Yves Porcheron
Don Powell
Frank Proctor
Cliff Ricketts
Dominique Rivière
Farouk Rizk
Bill Robinson
Brian Sadler
Mario Sforzini
Don Swift
Derek Taylor
Lee Teichthesen
Ian Thorpe
Prasad Verma

Balfour Beatty Transmission Division
Central Electricity Research Laboratories
Ceraver
Conference Internationale des Grands Reseaux Electriques
Doulton Insulators Ltd.
Dulmison Pty. Ltd.
Electric Power Research Institute
Lapp Insulator Co.
NGK Insulators Ltd.
Raychem UK Ltd.
Rosenthal (Hoechst Ceramtec)

Chapter 1

Introduction

1.1 Definitions

Insulators, for the purposes of this book, are the devices which are used on electricity supply networks to support, separate or contain conductors at high voltage. A special case, the insulating tools which are used in the maintenance of live apparatus, is included because of the many features in common with classical insulators.

Confusion often arises because of differences in American and English nomenclature. The English names are used here but a comparative table is given in the Appendix A.

1.1.2 Functions of insulators

All insulators have dual functions, mechanical and electrical, which commonly present conflicting demands to the designer. The most serious complicating factor is the impossibility, in practice, of providing an ideally nonconductive element. All insulators have external surfaces which will become contaminated to some extent in service. The contamination will carry leakage current: the surface layer, on a typically polluted insulator, will contain inert mineral matter, electronic-conductive dusts like carbon or metal oxides, soluble salts and water. This layer will behave as a highly variable and nonlinear resistor, in most cases unstable in the presence of electric fields. The leakage current which it carries will give rise to heat, electrochemical products of electrolysis and electrical discharges. Secondary consequences will range from electrochemical erosion through discharge ablation to complete by-passing of the electrical insulation by flashover.

Leakage current and its consequences largely govern the design of an insulator, especially one which is to be used outdoors in atmospheric wetting and pollution. A large transmission-type insulator (Fig. 1.1) presents a massive and impressive appearance: it is, however, not a monument of strength but a demonstration of weakness! The large bulk for a relatively trifling mechanical duty is imposed by the need to use a brittle insulant, while the complex and convoluted profile exists solely to combat the effects of leakage. Whereas a column of air, three metres long, is ample insulation for a 400 kV conductor, a surface liable to outdoor contamination must provide at least threefold that path length.

Fig. 1.1 *400 kV Housing – porcelain*
This is an unjointed, hollow cylinder, shaped by turning.

Another result of the interrelation between electrical, mechanical and environmental variables has proved to be the impracticability of completely designing an outdoor insulator on purely theoretical bases. Although the making and selection of insulators is no longer a 'black art' it is certainly not yet an exact science.

1.1.3 Classification of insulators

The principal classes of insulator are illustrated in Appendix A. Their main functions, as ties in tension, as struts in compression, as beams in bending and as containers in hoop-stress are usually supplemented by others: for example, the ABCB support (Fig. 1.2) is normally a strut and a container but will have superimposed bending, under fault conditions, because of electromechanical effects of the current in the supported metalwork.

1.2 History of insulators for transmission systems

1.2.1 Transmission lines

Insulators are much older than power transmission: telegraph insulators, introduced about 1835, had reached an advanced state of evolution by 1878[1], whereas the first transmission line was not run before 1882. This short line, from Miesbach to Munich, for 1343 V DC., was designed by von Miller and Duprez, to run an artificial waterfall at the Munich Electrical Exhibition.

Fig. 1.2 *Spray live washing of 400 kV, air-blast circuit breaker*
Porcelains are over 3 m in height

The next major steps were taken by C. E. L. Brown, son of the Charles Brown who founded the Brown Boveri Company. In his early twenties he first built a 2 kV line over the 8 km from Kriegstetten to Solothurn in 1886 and then went on, in 1891, to design a line no less than 175 km in length, for 15 kV originally but later for 25 kV. This ran from Lauffen, on the Neckar, to the Frankfurt Technical Exposition[2]. Brown used oil-bath insulators, based on the Johnson and Phillips patented telegraph insulator (Fig. 1.3).

Development was then rapid: a 40 kV line from Gromo to Nembro of 1903 was followed by others, rated between 50 kV and 66 kV, in Germany, France and Spain, all operating by 1910. As for North America, Lundquist[3] as early as 1912 describes lines for 140 kV in Michigan and illustrates a switch insulator for 150 kV. It is hard to believe that his pictures of transmission towers and insulators are more than seventy years of age, both metalwork and ceramics being barely distinguishable from current practice.

Transmission practice has diverged, over the last fifty years, into the long lines and very high voltages, as used in USA and the USSR, and the highly dense and interconnected networks of Europe. Voltages of 750 kV and above have been needed, as well as high-voltage direct current, for the former but only rarely for the latter. A second divergence has been between continental Europe and USA with UK. Germany and much of central Europe agree in favouring longrod porcelains to strings of discs and in a somewhat more receptive attitude to glass than that of USA and UK.

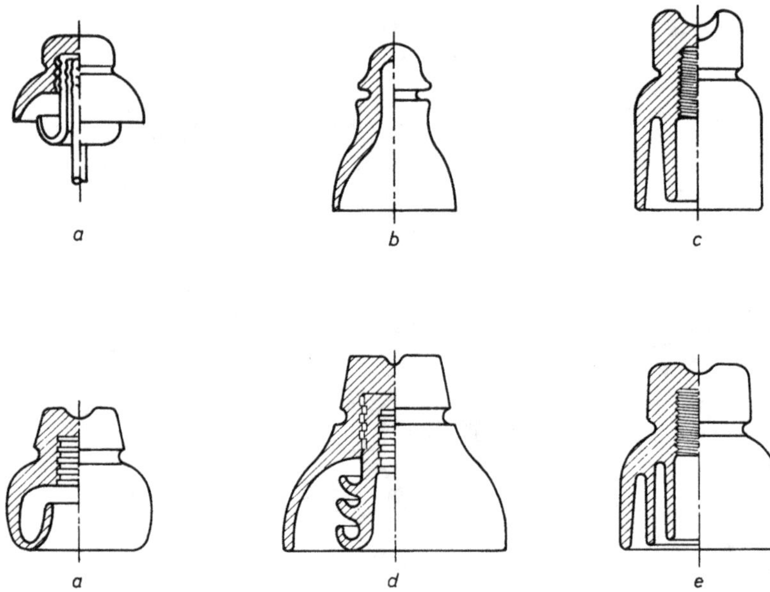

Fig. 1.3 *Telegraph and early transmission insulators*
a Johnson & Phillips, 1876
b Siemens, 1850
c Double-bell, 1858
d Multiple-chamber oil bath, 1891
e Three-skirt 10 kV pin, 1895

1.2.2 Insulator evolution: materials and shapes

Both glass and porcelain had been used before 1878 to insulate telegraph lines, while Gavey[1], in a far-sighted paper, also mentions and discusses accurately the limitations of polymers like ebonite and impregnated wood. All were thus natural choices for high-voltage lines and all were used, in various forms.

The electric and mechanical stresses which telegraph insulators were required to withstand were evidently negligible in comparison with those from power-line duty. The new demands soon disclosed serious shortcomings in both the materials and designs of insulator which were mere scaled-up copies of the telegraph types. In particular, puncture by electric stresses of ceramics which contained pores or flaws, cracks caused by differential thermal expansion or corrosive effects in metal or cement and flashover arising from dirt in combination with humidity, all showed the need for an evolution of design and manufacturing methods. This has been more of a succession of advances and retreats than an orderly progression, and continues an erratic course up to the present time.

Commerce has also had a great deal of influence on insulator evolution. Brent Mills[4] describes some of the ferocious battles in the USA which were fought over such details as ball-and-socket couplings between disc insulators and which hindered adoption of 'the best' in favour of what could be sold without patent infringement.

Some of the earliest nineteenth-century insulators (Fig. 1.3) are immediately recognisable as descendants of telegraph pin types, although the large sizes and weights of pin insulators for 66 kV, as used on the Molinar–Madrid line or even as high as 88 kV in North America, as recorded by Lundquist[3], led to great difficulties in manufacture, handling and mounting.

Early American pin insulators used built-up assemblies of relatively thin shells, made from dry-pressed porcelain, from ordinary annealed glass or sometimes from both in one unit. Some porcelain parts were glaze-jointed together; glass would generally be cemented-in, sometimes in the field rather than in the factory (with an easily comprehensible effect on quality). European porcelain technology was then in advance of American, enabling large single-piece or two-piece pin insulators to be made without multiple internal joints.

In all these insulators the ceramic or glass was loaded predominantly in compression, a mode in which these hard but brittle materials exhibit very high mechanical strengths, provided stress concentrations are avoided. For tensile-loaded insulators the earliest practice, again derived from telegraph lines, was to put the ceramic into compression between links. The basic egg-shaped insulator (Fig. 1.4), which survives today as the very strong guy-wire isolator, led the way, about 1907, to a disc insulator which also embodied interlaced links and compression loading, the Hewlett. This provided good creepage path between the electrodes, failed safe if smashed by impact in service and needed no internal cement. It was, however, quite difficult to make; in some designs it was subject to shattering by the freeezing of entrapped water and had poor properties as a generator of radio interference (RI). Although incapable of supporting really heavy loads, the Hewlett had a distinguished and long career, and is still in service in parts of the world where corrosion has defeated more modern designs.

Metal 'hoods' for insulators, introduced in the USA about 1903, allowed what were basically pin types to be stacked into posts (Fig. 1.4) and to resist considerable bending loads. This was a crucial step: it led on the one hand to the suspension-type cap-and-pin insulator, universally used ever since in hundreds of millions to support transmission lines, and on the other hand to the pedestal post and later the polyped, until recent times the dominant substation insulator and still the best in heavy-wetting conditions.

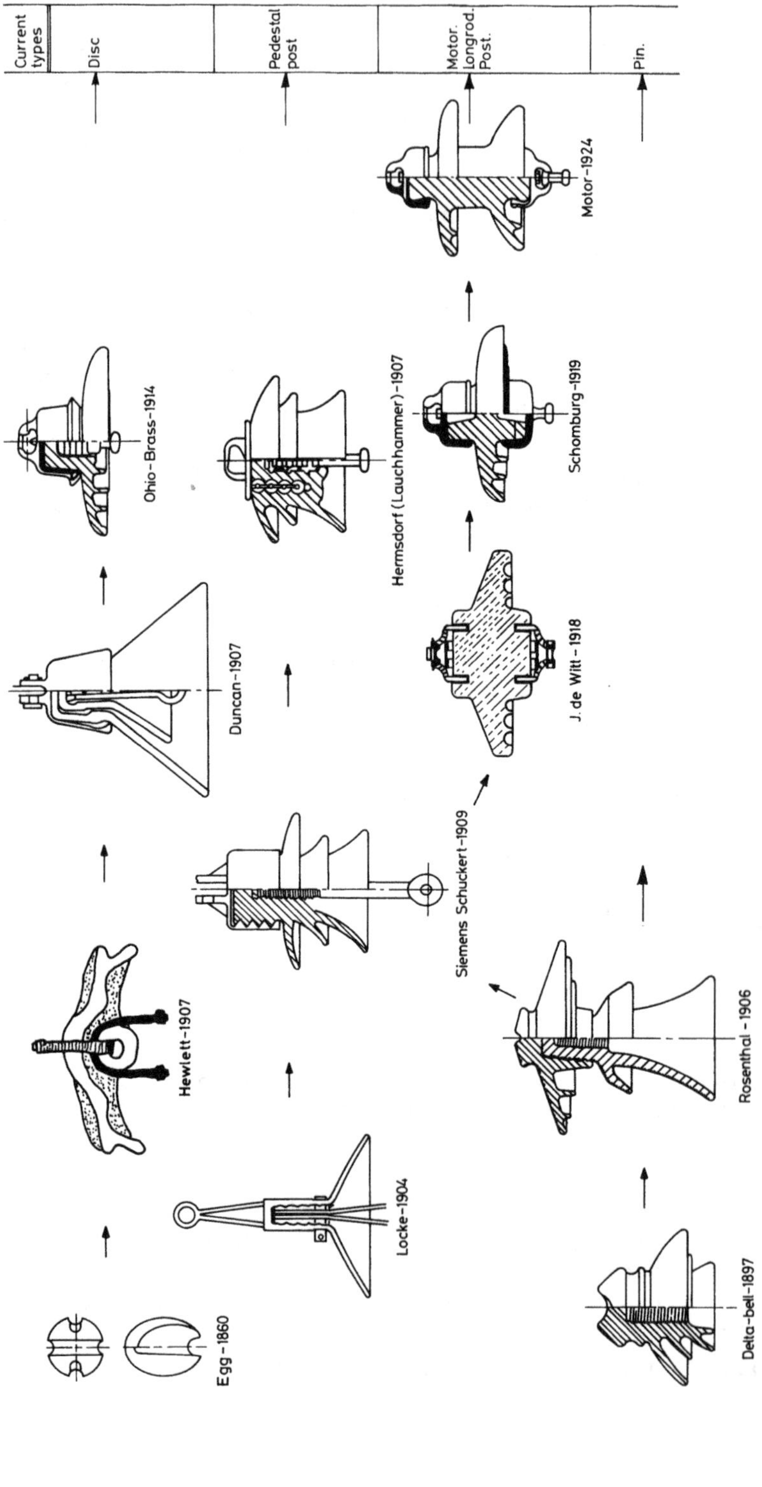

Fig. 1.4 *Evolution of insulators*

The unique feature of the cap-and-pin design is that it converts an applied tensile load into a radial compressive stress on the ceramic dielectric, which, as already stated, it withstands far more easily than tension.

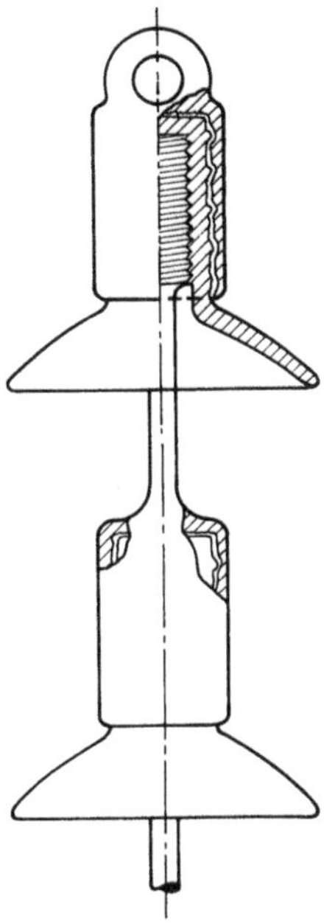

Fig. 1.5 *Early sectional strain insulator*
From Locke Insulator Catalogue, 1904. 'The porcelain and iron parts are threaded and engage so that they will not drop apart. . .'.
This insulator may never have been actually produced

Brent Mills illustrates a sectional strain insulator dated 1904 which is undoubtedly a cap-and-pin design, but this does not embody the radial compression system (Fig. 1.5). It appears that A. O. Austin, of the Ohio Brass Company, first employed radial compression and that his discs were of about 1907 vintage, although J. D. E. Duncan patented a multi-part cap-and-pin design in the same year.

Although the essence of this innovation was both simple and beautiful – the use of opposed conical surfaces to convert tension into radial compression by wedging action – a proper marriage of metal fittings, cement and ceramic dielectric turned out to be dificult of achievement, because of the hazards of practical service. In particular these were found to include effects of temperature and load cycling and corrosion.

The cement was seen as the hidden evil – it is fair to say that suspicions remain today[5] – and all manner of alternatives to its use were explored (Fig. 1.6). Some of these were fantastic, such as combining fusible metal with a spring ring. This particular design was used extensively on the 132 kV transmission grid, in Great

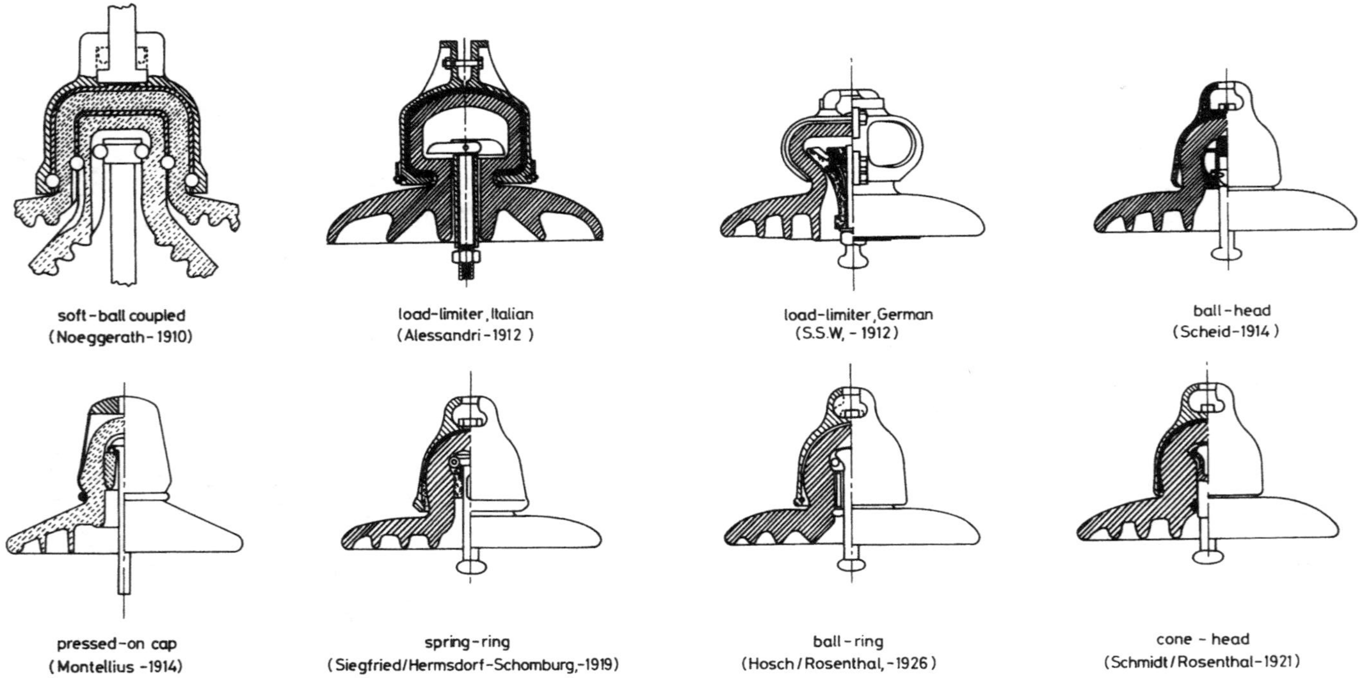

Fig. 1.6 *Some alternatives to cemented pins*
Several of these designs, from 1910 to 1935, were produced in commercial quantities. None was successful

Britain, and behaved disastrously. Another, the Nöggerath ball-ring type patented in 1910, was equally unsuccessful elsewhere. Probably these commercial defeats combined with the very high standards of porcelain manufacture, which had evolved in Germany, to encourage the direct use of porcelain in tension, in continental Europe.

R. M. Johnston, of the American Jeffrey DeWitt Company, had succeeded by 1918 in combining the worst features of buried metal and tensile-loaded ceramic, but Continental practice, first with the Motor insulator, marketed by Motor Columbus, and later with the Langstab or longrod, was to use two external caps, either cemented or held by poured metal, to apply tensile load to the porcelain body[6] (Fig. 1.4).

This move largely eliminated the corrosion failures, since the sole effect of cement growth or metal reaction was to put the porcelain into increased radial compression. Unfortunately, however, it also sacrificed the outstanding virtue of all cap-and-pin designs, their ability to maintain mechanical integrity even when shattered.

The longrod was the inspiration for the hollow-core post and later the line post and solid posts, both of light construction for overhead use, and of heavy construction for application in substations. From the longrod and line post also evolved the polymeric insulator, based on cores of resin-bonded glass fibre.

With the gradual passage to higher and higher voltages and with the introduction of nuclear generation, the penalties in direct lost revenue, and even damage to plant, arising from insulator flashover increased sharply. More attention was paid to the improvement of performance of insulators in contamination. Consequences of this change of emphasis towards increased electrical reliability included the evolution of strong disc insulators of exceptional creepage length and of profiles, for large substation insulators, which would enable these types to perform comparably to overhead-line designs. The Multicone or multiple-cone post used assemblies of ceramic interlocking parts to produce strong posts having much more creepage length than could conveniently come from conventional technique, like turning or sticking-up of jollied parts (see Chapters 2 and 3). This innovation again found little favour in North America and, in fact, has recently been criticised in Europe for mechanical shortcomings, when aged. Although porcelain multiple cones were introduced, in France, as long ago as the 1960s, their apparently obvious extension to glass did not occur before 1970, and then only on a limited scale.

It was not much before 1962, following the recognition that insulator performance could be predicted from laboratory tests, that shapes were first examined scientifically, and compared in reproducible orders of merit. The primary agent was the Salt Fog Test[7], one consequence of which was the abandonment of several fanciful shapes, including helically shedded posts and discs, which had been designed on false physical assumptions and supported by invalid tests like the artificial-rain procedure. The resistive-glazed or stabilised insulator was introduced during this period and gave an excellent performance, both in artificial and real pollution. Corrosion problems have limited its application since[8,9,10].

With increasing transmission voltages and numbers of subconductors in bundles, the mechanical loads to be suppported by overhead-line types rose above the 40 tonne level, which is about as much as can be provided by a single porcelain disc of a weight which can be handled conveniently. Multiple strings were introduced, e.g. in the UK, on a large scale, but these had the disadvantages of obtrusive appearance and mechanical complexity. Toughened glass discs, introduced by the Pilkington Company[11,12] in the UK in 1935, offered some amelioration because of their better strength for a given weight and size of hardware, but were evidently incapable of simple extrapolation to the 100 tonne levels which would be required for million-volt service. This provided logical entrée for the polymeric insulator, using the fibrous composite materials to give high ratios of tensile strength to both size and weight.

The problems of reliability proved, and have continued to prove, themselves difficult of solution: the matter is dealt with in Chapters 5 and 6, where polymer technology is described in detail. The virtues of polymeric longrods, however, and especially their small size and resistance to attack by vandals, led to their introduction at quite low voltages for which they really had little intrinsic applicability. It now appears that, for many of the duties where extremely high ratios of strength to weight are not necessary, it is advantageous to use a Hybrid principle, in which polymeric material is used for good surface properties and compactness but the fibrous composite core is replaced by ceramic[13].

The recent history of insulator development has been dominated by increases in size, rather than revolutions in technology, to meet the demands of million-volt and direct-current transmission. Disc insulators have now been made with a mechanical capability of 78 tonnes and hollow substation types of up to 8 m in height. It is curious to reflect that whereas the thermionic valve and its associated engineering has been swept away, in little more than a decade, by solid-state devices, no such drive into obsolescence has been suffered by ceramic insulators, by competition from polymers, fibres or any other new materials.

1.3 Critical elements of an insulator

The considerations which decide how an insulator shall be made up and how it will perform in service are three: the properties of the materials within it, its ability to operate in adverse weather and contamination and its cost. Details of these matters appear in subsequent Chapters: a summary of the essentials is given here.

1.3.1 Properties of materials for insulators

The three main materials in any insulator are the dielectric, the terminations which couple the dielectric to the mechanical structure and the intermediaries such as cements, lubricants or paints.

The dielectrics, commonly porcelain, glass or polymer, are required to hold off the applied potential difference, either power frequency or DC., typically for several decades without failure. They must also resist impulsive voltages arising from lightning or switching operations without puncture. Since the surfaces of the dielectrics will, in service, always be electrically conductive to some degree because of humidity and dirt, a good resistance to electrical discharges and electrochemical products, as well as to normal corrosion and ablation, is also essential.

From the summary of properties (Table 1.1) it is seen that there is a wide divergence of values between different dielectrics and that the most commonly used material, electrical porcelain, has quite poor electric strength. This apparently suprising weakness is, however, an unreal one since, even in suspension insulators where the dielectric stress is higher than in any other type, long-term values higher than a few kV/mm are unknown. Failure under power frequency may have been common in early porcelain insulators where the dielectric was both porous and mechanically overstressed, but is currently rare, except where incipient failure has already arisen from mechanical, thermal or corrosive forces.

The life expectancy of dielectrics is subject to controversy. Early glass and porcelain bodies usually failed for obvious reasons connected with poor manufacture, such as porosity or the presence of inclusions. With improved technology and also with increasing mechanical duty, however, failures were noted which could not be blamed on defects in materials since these had substantially been eliminated.

Glass dielectrics are self-indicating of failure because any puncture or crack results in visible shattering of the whole body. Porcelain, on the other hand, may be punctured internally with no external indication – a most serious operational shortcoming since factors of safety are lowered insidiously – and thus its failure rate obscured until testing is made. It was firmly established from the earliest days of glass that a 'spontaneous' shattering in service of some 0·1% annually could be expected on AC. lines, where toughened discs were used[14]. Much later observations, especially from the Canadian 735 kV lines employing porcelain which was highly stressed both electrically and mechanically, suggested that equal or even higher failure rates were occurring, in comparison with glass[15].

Work in England during recent years has shown conclusively that the failure rate of porcelain disc insulators in a 400 kV string is correlated with the position in the string, and thus with the electric stress on the unit. Although ancillary effects, like cement growth and pin corrosion, would also be expected to become worse with increased electric stress, the possibility remains that the dielectric does age and that the rate is electric-stress dependent. Effects of piezoelectric motions within the dielectric have been invoked as a potential failure mechanism; it seems probable, however, that mobile ions, of alkali or alkaline-earth

Table 1.1 *Summary of properties of insulator dielectrics*

Property	Unit	Glazed porcelain	Toughened glass	Polymer	RBGF
Density	g/cm	2·3–3·9	2·5	0·9–2·5	2·1–2·2
Tensile strength	MPa	30–100	100–120	20–35	1300–1600
	$lb/in^2 (\times 10^3)$	4·0–14·0	14·5–17·4	3.0–13	190–230
Compressive strength	MPa	240–820	210–300	80–170	700–750
	$lb/in^2 (\times 10^3)$	34–120	30–40	11–24	100–107
Tensile modulus	GPa	50–100	72	0·6–16	43–60
	$lb/in^2 (\times 10^6)$	7–14	10·1	0·1–2	6·0–8·0
Thermal conductivity	W/m° K	1–4	1·0	0·17–0·9	0·2–1·2
Expansibility (20–100 °C)	$(\times 10^{-6})$/K	3·5–9·1	8·0–9·5	45–200	7·5–20
Permittivity (50–60 Hz)	Air = 1	5·0–7·5	7·3	2·3–5·5	2·5–6·5
Loss tangent (50–60 Hz)	$(\times 10^{-3})$	20–40	15–50	0·1–5·0	5·0–20
Puncture strength	kV/mm	10–20	> 25	> 25	3·0–20
Volume resistivity (at 20°C)	Ωcm	10^{11}–10^{13}	10^{12}	10^{15}–10^{17}	10^{11}–10^{14}

species, are the principal culprits, since glass also shows a similar aging. It is well established that both glass and porcelain rise rapidly in electrical conductivity with temperature and that failure rates are also temperature dependent, especially under direct voltage where migration would be unidirectional.

It must be the case that the internal microcracks in ceramic dielectrics, associated with the granular structure of porcelain and with the inevitable inclusions in moulded glass, will develop with time and weaken the dielectric. There seems no obvious way in which this process could be accelerated by electric fields of moderate intensity, however, to explain the observed correlation with voltage-distribution.

The fibrous core of a polymeric insulator is a special case. Its mechanical lifetime is limited by statistical processes[16,17], and sometimes curtailed by invasion. As with classical dielectrics, however, failure is accelerated by increased mechanical or electrical stress. The terminations are invariably of metal, commonly malleable iron castings which are heat-treated and galvanised. Non-ferrous alloys based on copper are used for some, especially electric traction, applications while aluminium forgings and zinc-alloy diecastings are sufficiently strong for light duties and low voltages. Embedded pins or bolts are usually of steel, occasionally a stainless alloy, and common practice, at least on DC, is to use a sacrificial zinc sleeve to combat corrosive growth of the buried part. Corrosion reduction is a complex matter with insulator terminations because of the heavy ionic depositions which occur when wet layers of pollution are electrolysed by leakage current.

The intermediaries which couple the dielectric to the metalwork are commonly cement, either Portland or aluminous, or rarely fusible sulphur-based adhesives. Low-melting lead–antimony alloys have found application in a few longrod types. Thermoset resins are widely used for polymeric insulators. Both corrosion protection and some degree of lubrication are needed in some types, particularly cap-and-pin suspension discs. These are provided by bituminous layers, the thickness and properties of which have large effects on the mechanical behaviour, in particular the tensile failing load and the toleration of thermal cycling, exhibited by cap-and-pin discs.

1.3.2 Operation in adverse conditions

The ability to function as an insulator, i.e. to resist flashover, in adverse conditions is primarily determined by the profile of the dielectric, but secondarily by the attitude of the insulator (vertical, inclined, horizontal) and by the properties of the surface (hydrophobicity, roughness).

A fallacy which has dogged insulator design from the earliest days is that water is the main enemy. Because the surface resistivity of glass and porcelain falls rapidly with increasing ambient humidity (Fig. 1.7) wet testing was introduced to give an order of merit for telegraph insulators, for which, indeed, low attenuation by parallel resistance is important. Wet testing led, however, to artificial rain testing and to the concept of 'protected creepage', both of which have had an adverse influence on the evolution of good shapes of insulator.

The fundamentals which decide whether or not a given insulator will operate when it is both contaminated and wet are the surface electric gradient along the creepage path between the electrodes and the internal electric gradient, in the arcs which span dry parts of the creepage path (Fig. 11.1). Flashover is inhibited by a low gradient in the creepage path and a high gradient in the arcs. Since arc gradients fall with increasing current, any steps which reduce leakage will also improve flashover voltage.

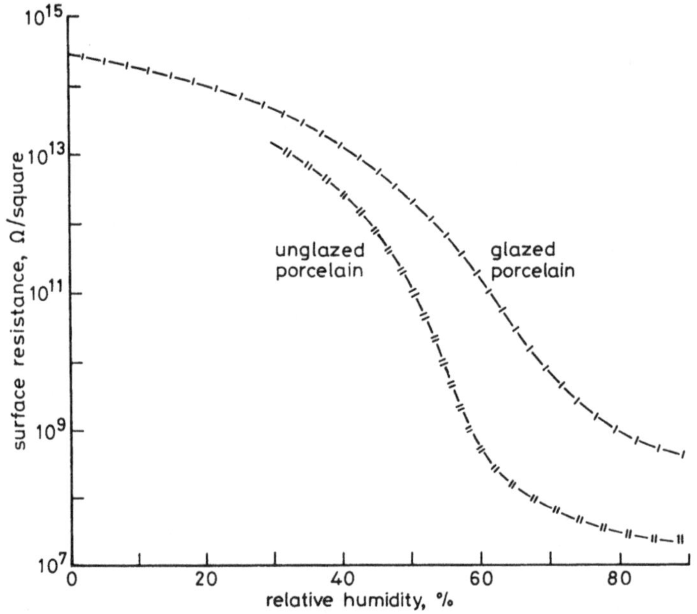

Fig. 1.7 *Surface-resistivity variation with humidity (after Reference 43)*
These curves are from low-tension-grade material: variations are less, with some high-tension grades

The guiding principles in design of insulator profiles are therefore as follows: maximize the creepage path without spoiling the aerodynamics of the surface, since the principal contaminating mechanism is deposition from air flow of suspended dust or droplets; promote drainage of surfaces by avoiding near horizontal flat areas; maintain airspeeds over the insulator by avoidance of vortex generators, which deposit dirt in the convoluted parts of the surface by cyclone action; increase surface resistance by reducing the adhesive forces and promoting breakup of wet films into isolated drops.

In recent times, it has been recognised that there is no such thing as a 'good' shape of insulator, in absolute terms, since the performance depends on the rate of wetting. An illustration is the behaviour of an inclined, long-creepage insulator, respectively, under heavy contamination with light wetting, as in a coastal desert area, and in light contamination but heavy wetting, as under artificial live washing. Close-packed skirts are good for the first case but bad for the second. An open profile or one which avoids short-circuiting of the creepage by cascades of water is good for the heavy-wetting case but may be poor for the first, light-wetting condition.

The implications of this principle – that 'best shape' depends on wetting-rate – are wide, extending to the validity of artificial tests and the efficacy of surface treatments against flashover. Detailed comments are made in the relevant Chapters.

An 'adverse condition' which is presenting increasing difficulties for users of insulators in some countries is attack, by vandals or shooters, using missiles, or impact, from rough handling or because of careless erectors. Measures to absorb kinetic energy are therefore having to be designed into insulators which rely on brittle dielectrics.

1.3.3 Costs of insulators

The purchase price of insulators, especially those of glass or porcelain, is totally dominated by market forces and bears no relation either to the technical importance of insulation in a transmission line or to the cost of replacement and revenue loss, in case of failure. In a word: insulators are absurdly cheap.

The best illustration is historical. A. O. Austin, quoted by Lundquist in 1912[3], shows porcelain insulators of different types as selling at US cents 6·25 per pound during a period when the cost of a linesman was US $2·50 per day. These were primitive insulators, probably assembled in the field, whose reliability was assessed by A. S. Watts[18] in 1910 as low: 'A thoroughly practical and successful transmission insulator is not yet on the market'. Yet, in the mid-1980s, insulators which are required to meet stringent requirements of reliability, freedom from radio-interference generation, mechanical, thermal and electrical duty are sold for little more, per pound, than in 1912. Assuming a typical day rate of US $80·00 for the 1980's linesman, the effective insulator price has fallen by a factor greater than 30!

Although the raw materials for glass and porcelain insulators are very cheap, averaging no more than US cents 6·00 per pound, the high processs temperatures are expensive in energy and the handling and labour costs, of some 20 production steps, are considerable. The low selling prices have therefore led to starvation of funds for technical research and development, among the older established manufacturers, since gross profits have fallen so low.

The emergence of insulator production in parts of the world where labour costs are much lower than in the industrial North has further increased the downwards pressure on purchase price and upon the level of quality. A change of attitude towards insulators is now becoming apparent, however, in some supply utilities. They start to consider true costs, rather than mere purchase prices, when ordering new lines. They first calculate the maintenance and treatment costs, such as live washing or protective greasing, the prospective costs of unplanned outage caused by insulator failure (amounting to some US $\$10^5$ per day, if the lost line is an important one), and the costs of replacing the insulator at the end of its useful life. When they further consider that the component of installed cost of a typical transmission line which is represented by the insulators is usually 7% or less, the economic case for better insulators, even at several times current prices, is seen to be made.

The economics of polymeric insulators, much different from those of ceramic ones, have contributed to the change of thought. Typical material costs, of good-quality fibrous composite cores or of track-resistant elastomers used for outer housings, are likely to be at least tens and probably hundreds of times higher than for ceramics. Nevertheless, users have found that their savings from ancillary advantages like vandal resistance and ease of handling, as well as the major matters of reliability and lifetime, mentioned above, have often justified very much higher purchase prices. A curious and interesting case is that of insulators for installation in the Arctic or in inaccessible mountain regions. Here the helicopter delivery costs are such that the weight is the only serious property to be considered, price becoming largely irrelevant!

Once the designer and inventor are free to take account of the remarkable materials which are now available, without inhibition by the need to compete on purchase price with conventional ones, much needed technical advances may be expected (Chapter 17).

Chapter 2

Insulating materials

2.1 Basic nature of insulator dielectrics

The principal dielectrics which are used in insulators are ceramics and polymers. Both are built from 4-valent atoms, silicon in the case of ceramics and carbon in polymers, which have the capability of forming extended structures.

Silicon combines with oxygen to form either a crystalline mineral, quartz, the structure of which is temperature-dependent but always based on Si–O tetrahedra with all corners shared (Fig. 2.1), or a glassy form, fused silica. When other elements are added to the fundamental two an enormous range of silicates, either crystalline or glassy, results. Electrical porcelain (Fig. 2.2) is made up of a glassy matrix containing crystals of various kinds and grains of the original component minerals, some of which are partly dissolved. Electrical glass ideally is a mixture of silicates in vitreous form without inclusions. In practice, it always contains some gaseous bubbles and fragments of refractory material as imperfections. Porcelain insulators are always glazed.

Carbon combines with itself as well as with hydrogen, oxygen and many other elements to form chains and rings which can be made to cross-link into networks of great variety and complexity (Fig. 2.1). Pure polymers, containing only a single molecular species, are rarely used in insulator manufacture, the universal practice being to employ copolymers and mineral fillers for the achievement of the required mechanical and electrical properties.

The salient and fundamental difference is in stability. The strong electrostatic bonds between silicon and oxygen which hold the ceramics – porcelain and glass – together confer high melting point, mechanical strength (but also brittleness) and resistance to chemical attack.

They also contribute to a high value of surface free energy, the thermodynamic quantity which decides the 'stickiness' or strength of adhesion to contaminants, and hence causes ceramic insulators to be readily wettable and easily polluted.

The polymers, on the other hand, are weakly bonded on the molecular scale. They are all decomposed by heat at a few hundreds of degrees Centrigrade,

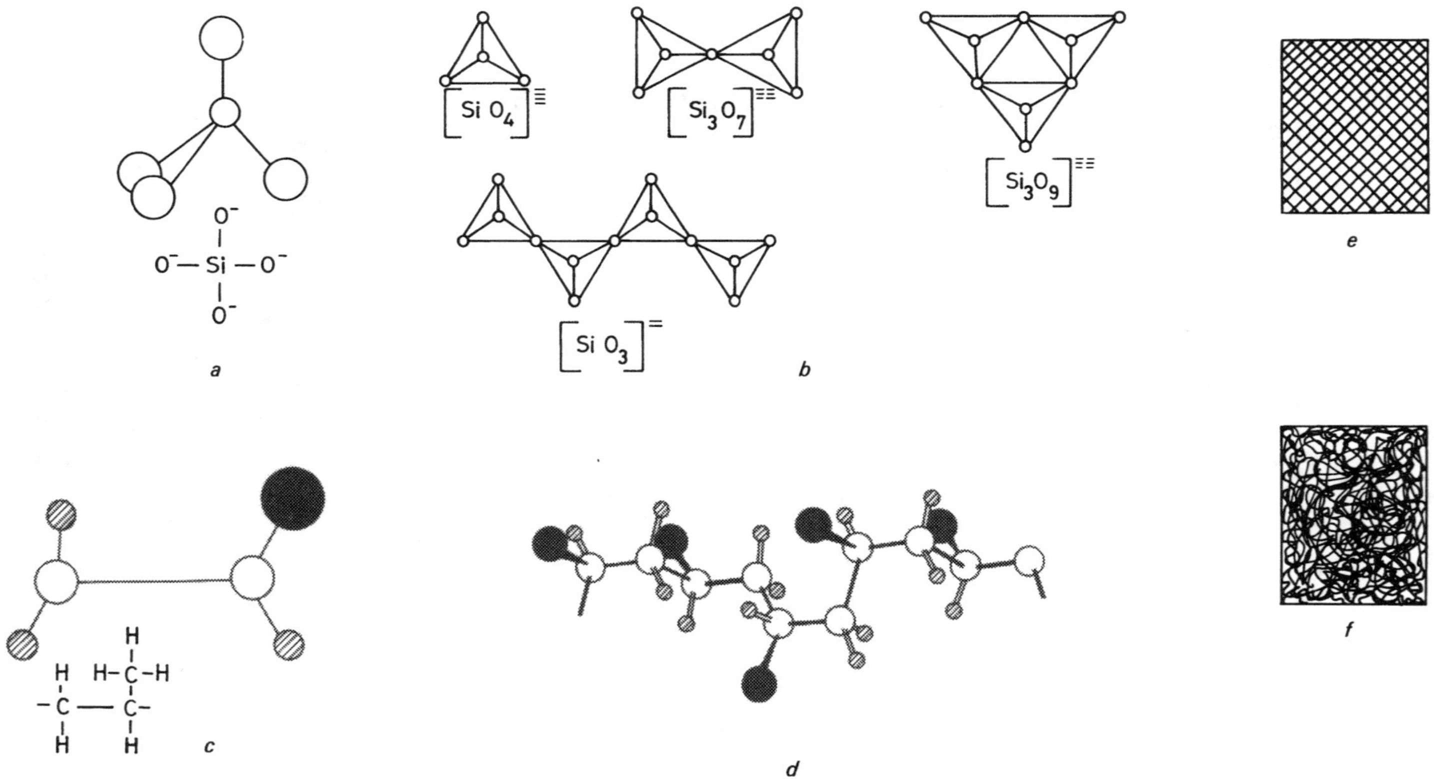

Fig. 2.1 *Silica and carbon as building blocks*

a Exploded view of SiO_4 tetrahedron
b Some modes of linkage in silica and silicates
c Element of polyolefine molecule (polypropylene)
d Form of molecule is helical on triangular section (polypropylene)
e Glassy or crystalline surface is rich in bonds and is 'adhesive'
f Polyolefine has 'ball of wool' make-up with saturated surface and few bonds

many are subject to surface damage even by the quanta of ultra-violet light from sunshine, and all are capable of reaction with atmospheric oxygen under working conditions which generate electrical discharges. Their most serious defect is that their foundation element, carbon, is a good electrical conductor in most of its uncombined forms. Surface attack of some kinds, especially by discharges at high temperatures, therefore produces conducting paths or tracks which may cause flashover and eventually destroy the insulator.

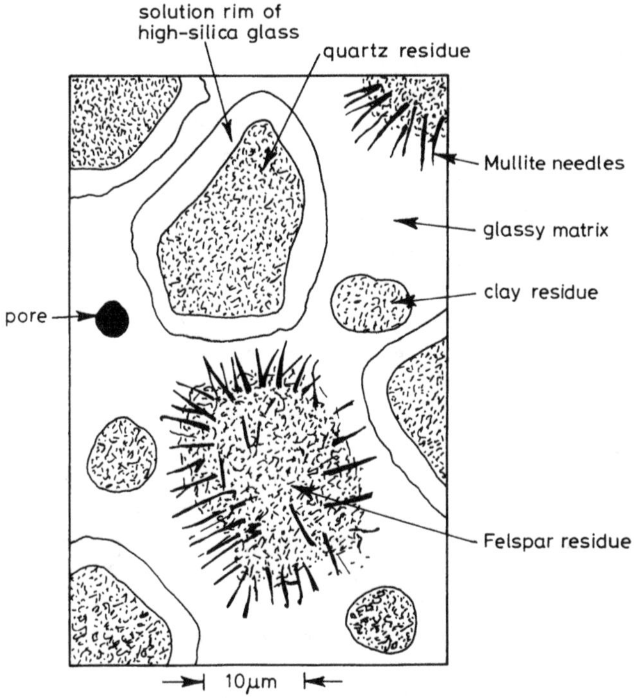

Fig. 2.2 *Electrical porcelain – diagrammatic representation*
The mineral residues of quartz, felspar and clay, as well as the needle crystals of Mullite, have very high volume resistivities. The glassy matrix is the principal electrical conductor

Apart from the resilience and total absence of brittleness which are conferred by the fibrous molecular makeup of polymers, their outstanding advantage is in their low values of surface free energy. Thermodynamically, the outer skin of hydrogen atoms, bonded to carbon skeletons, which most polymers present to the atmosphere enables them to resist wetting and contamination to a far greater degree than is possible for ceramics.

From the earliest insulator used by Brown in 1891, which embodied reservoirs of hydrocarbon oil, the desirability has been recognised of taking advantage of the good qualities of ceramics while reducing their surface wettability and attraction for dirt. The search for utilising this principle continues even today.

2.2 Properties of electrical porcelain

2.2.1 The determinants

The properties of a piece of electrical porcelain are largely determined by the demands of the manufacturing process and are not easy to predict *a priori*. The piece is a sintered aggregate comprising particles of different sizes, crystals and pores. This aggregate is held together by a glassy matrix which has an entirely different glassy skin, the glaze, on its surface. The principal chemical and physical features are determined by the raw materials and firing cycle, but the details are heavily influenced by the complexities of the manufacturing process.

Details of a typical sequence leading to the finished piece are given in Chapter 3, but the main steps, in wet process manufacture, are blending of clays and minerals in water, removal of the water, formation of the shape, glazing and firing, sometimes followed by refiring if the desired properties are not realised at the first attempt. Each step imposes its own constraints.

We consider only one property, plasticity. During manufacture the wet material must tolerate gross deformation in an extruder, must flow into complex moulds during pressing, retain its shape while drying; in other steps it must submit to turning in a lathe. Plasticity is essential for these operations; it is conferred by a high proportion of small particles in the mix (Table 2.1). However, too high a proportion of very small particles, in the fully wet or slurry stage, makes the filter-pressing time, when slurry is converted to semi-solid cake or slab, uneconomically long. Thus the particle-size distribution in the finished piece, on which the properties, especially mechanical, depend very heavily, is not a matter of predetermination but of manufacturing compromise. The designer gets what can be made at reasonable cost and not necessarily what he would like!

Another determinant of the properties of porcelain is shrinkage. Large changes in dimensions occur as the water is progressively removed, before the piece is fired: drying therefore has to be done carefully if internal cracks are to be avoided, especially in large pieces. As it goes into the kiln an unfired or 'green' piece may contain 35% of open pores, but this proportion commonly falls to 7% or less during firing. Large changes in linear dimensions therefore occur, of 3% or more, as the piece becomes more dense. These are accompanied by chemical dehydration, chemical reactions, crystallisation from glassy phases, sintering and mingling of glaze with substrate.

The final properties depend on the microscopic structure within the body and the macroscopic compression, in the external glaze, which is produced by deliberate thermal mismatch. Pore-size distribution and particle-size distribution, as well as the presence or absence of internal micro-cracks, largely determine the mechanical performance and the electrical puncture strength.

Generally, the smaller the pores, crystals and residual particles in the final body the greater its intrinsic mechanical strength (Fig. 2.3). Also generally, the substitution of alumina for quartz improves strength by reducing internal thermal mismatches between residual particles and glassy matrix. However,

Table 2.1 *Particle size distributions in clays and bodies*

Material (μm)	< 0·05	0·05–0.10	0·10–0·25	0·25–0·50	0·50–0·75	0·75–1·00	1·00–2·00	> 2·00
Ball clay (% in range)	20	23	38	12	7	0	0	0
Kaolin (% in range)	0	0	5	8	21	21	66	
Siliceous body (% in range)	7	7	9	7	4	3	11	52

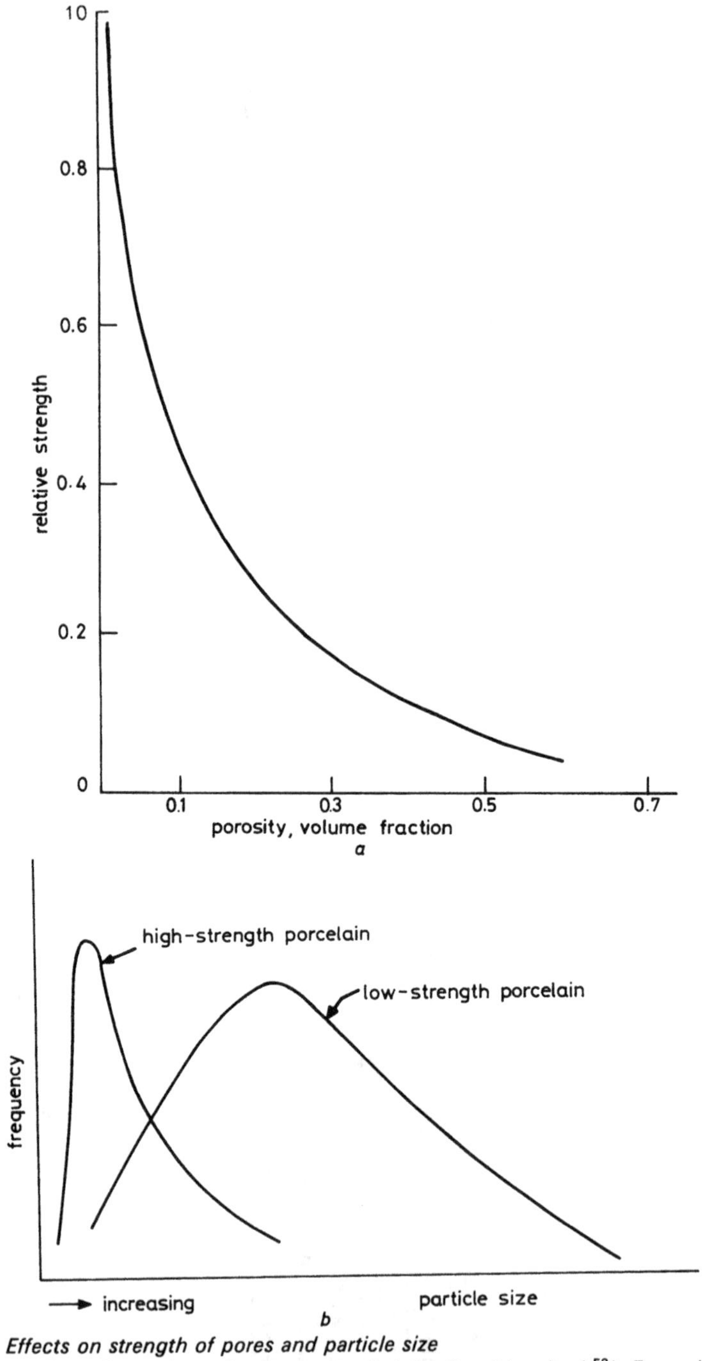

Fig. 2.3 *Effects on strength of pores and particle size*

a Effect of porosity on fracture strength (after Ryschkewitsch[53]). For a given porosity, smaller pores give higher strength

b Qualitative particle-size distributions in high-strength and low-strength porcelains based on dry-ground alumina. (after Reference 58)

since surface defects rather than internal ones are the more dangerous sources of fracture cracks, deliberate enhancement of a body's expansibility, by the incorporation of certain crystalline forms of silica, is sometimes employed to increase the glaze mismatch, and therefore the macroscopic surface strengthening (Fig. 2.4).

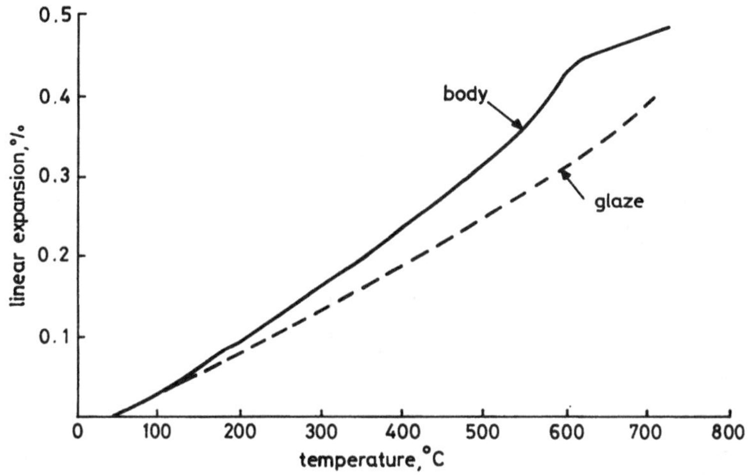

Fig. 2.4 *Mismatched expansibilities: body and glaze*
Once the glaze is 'frozen', contraction of the body puts it into compression. Final glaze stress may exceed 90 MPa

2.2.2 Mechanical properties

The mechanical properties of representative electrical porcelains (Table 2.2) require cautious treatment. Because of the brittle nature of the material, apparent tensile strengths are not normally distributed, but follow a skewed characteristic (Fig. 2.5), to which a good fit can often be obtained by application of Weibull statistics[19,20]. The practical consequences are that mechanical strengths, as determined by breaking cylindrical rods of different porcelain formulations, are unreliable guides to the comparative strengths of finished insulators, and that factors of safety which are based on normal distributions turn out, in practice, to be optimistic.

Where tensile failure may be a highly unwelcome event for the insulator user, for example in a longrod which drops the conductor on fracture, it is universal practice to proof-test each piece to an agreed fraction of ultimate load, thus rejecting the 'rogue' pieces (Fig. 2.5). Even in cap-and-pin designs, which are free of this objection, some manufacturers overdesign the porcelain so that the statistics revert to those of the metal fittings. These, the weakest links, have narrow distributions of failure because they are metal, having ductility and intrinsic crack-stopping properties.

Table 2.2 *Mechanical properties of siliceous and aluminous porcelains and of glass*

Property	Siliceous Porcelain		Aluminous Porcelain		Toughened glass: Alkali lime silica
Density (g/cm^3)					
Bulk	2·26–2·42		2·60–3·25		2·30–2·60
True, without pores	2·42–2·50		2·78–3·47		
Unglazed/Glazed	U	G	U	G	
Strength (MPa)					
Flexural	42–90	56–120	100–140	120–170	200–250
Tensile	21–42	28–56	50–70	60–80	100–120
Compressive	280–450	380–690	400–600	500–700	700
Fracture impact energy (J)	2·0–3·0		2·5–4·0		5·0–6·0
Modulus elastic, tensile (GPa)	55–80		80–120		60–70
Expansibility (20–100°C) ($\times 10^{-6}$/°K)	3·5–5·5		4·6–6·0		8·0–9·5
Thermal conductivity (W/m°K)	1·0–2·5		2·0–25·0		0·5–0·9
Specific heat (20–100°C) (J/g°K)	0·46–0·72		0·11–0·13		0·5–0·67

2.2.3 Electrical properties

The electrical properties which are generally specified (Table 2.3) for electrical porcelain intended for insulator dielectrics are puncture strength, permittivity, loss tangent, volume resistivity and T_e (temperature, in deg. C, for volume

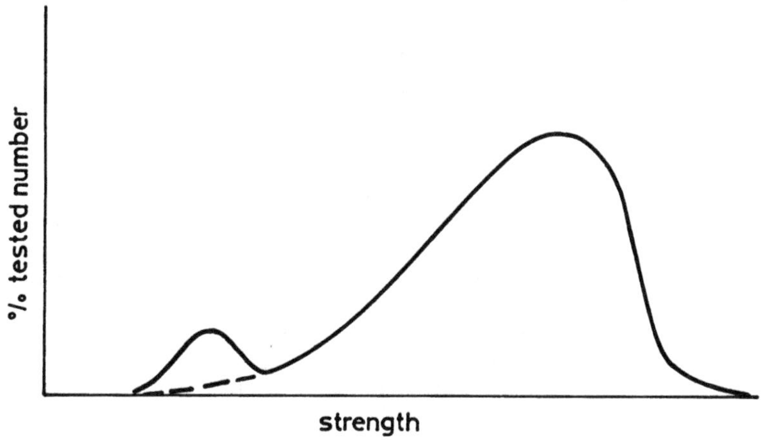

Fig. 2.5 *Skewed strength distribution of ceramic pieces*
The minor peak represents defective pieces. The skew is fundamental, arising from statistical strength dependence both on stress and tested volume (after Taylor[89] and Weibull[19])

resistivity = $10^6\,\Omega$ cm). In very few practical cases are the electrical properties critically important for the successful operation of the insulator, since the dimensions are usually determined by the mechanical and thermal requirements. For example, the thickness of a porcelain disc insulator in the head is commonly about 20 mm, which leads to electric stresses, under AC energisation, no higher than 1·5 kV/mm. Ordinary electrical porcelains have puncture strengths at least 5 times higher than this. Electrical pressure testing is therefore practised in production solely to seek out pores, flaws or cracks in the material which might lower the mechanical strength of the insulator but which are not visually evident. It is common practice to test disc insulators at voltages high enough to produce intermittent external sparkover – 80 kV or more – and to apply similarly severe stresses radially through the walls of hollow posts, especially when these are to be used as pressure vessels in switchgear[21].

Puncture strength under impulses is, however, relevant because of lightning strikes to power lines, and to a lesser extent because of surges generated by switching operations. A fast-rising transient voltage may exceed the puncture strength of the porcelain, within the head of a disc insulator or between the metal parts of a pedestal post, in a time too short to allow external sparkover to develop. A path then exists for subsequent penetration by water and a source of mechanical cracking is created, in both cases without visible evidence. Measurements made as long ago as 1941[22] showed the poor performance of porcelain, relative to other common dielectrics, and also the considerable role of surface defects.

Table 2.3 *Electrical properties of siliceous and aluminous porcelains and of glass*

Property	Unit	Siliceous porcelain	Aluminous porcelain	Toughened glass: Alkali lime silica
Permittivity				
(50–60 Hz, 20°C)	air = 1	5·0–6·5	6·0–7·5	7·3–7·5
(1 MHz, 20°C)	air = 1	4·8–5·6	5·0–6·5	7·1–7·5
Loss tangent				
(50–60 Hz, 20°C)	$\times 10^{-3}$	10·0–25·0	12·0–30·0	15·0–60·0
(1 MHz, 20°C)	$\times 10^{-3}$	5·0–12·0	5·0–12·0	5·0–12·0
Puncture strength				
(50–60 Hz, 20°C)	kV/mm	10·0–20·0	10·0–20·0	> 25·0
Impulse puncture strength				
(1/5 μs)	kV/mm	40·0–50·0	40·0–50·0	170·0–220·0
$\varrho \equiv$ Volume resistivity				
20°C	Ω cm	10^{13}	10^{12}	10^{12}
300°C	Ω cm	10^{6}	10^{11}*	$10^{5}-10^{6}$
$T_e \equiv$ Temperature for $\varrho = 10^{6}\,\Omega$ cm	°C	280–340	830–1070*	270–400

* Values fall rapidly with increasing per cent Na

The observed values of puncture strength, permittivity, loss tangent and volume resistivity, and the variations of these quantities with temperature, are consistent with the model (Fig. 2.6) of a glassy matrix, containing relatively mobile ions, which cements together an inert structure of grains, crystals and pores. An all-glass sample would have a much higher puncture strength since its internal electric-stress distribution would not be distorted by discontinuities of lower permittivity (glass/silica/air: 7·5/4·2/1·0). The permittivity is similarly diluted by the inclusions, from 7·5 to about 6·0, in relative numbers. Loss tangent and electrical conductivity are both dominated by ionic motions in the glassy matrix. Since ionic mobility rises rapidly with temperature, both quantities similarly rise, while replacement of sodium by the larger and less mobile potassium ions duly reduces loss and conductivity and raises the puncture strength.

The practical consequences of ionic motion are mainly seen in accelerated failure under DC stress compared with AC. The change of dimensions of ions, especially Na^+, on discharge at electrodes must introduce mechanical stresses there, while internal crack propagation may well be favoured by electrostatic forces at the surfaces of inclusions[22–26].

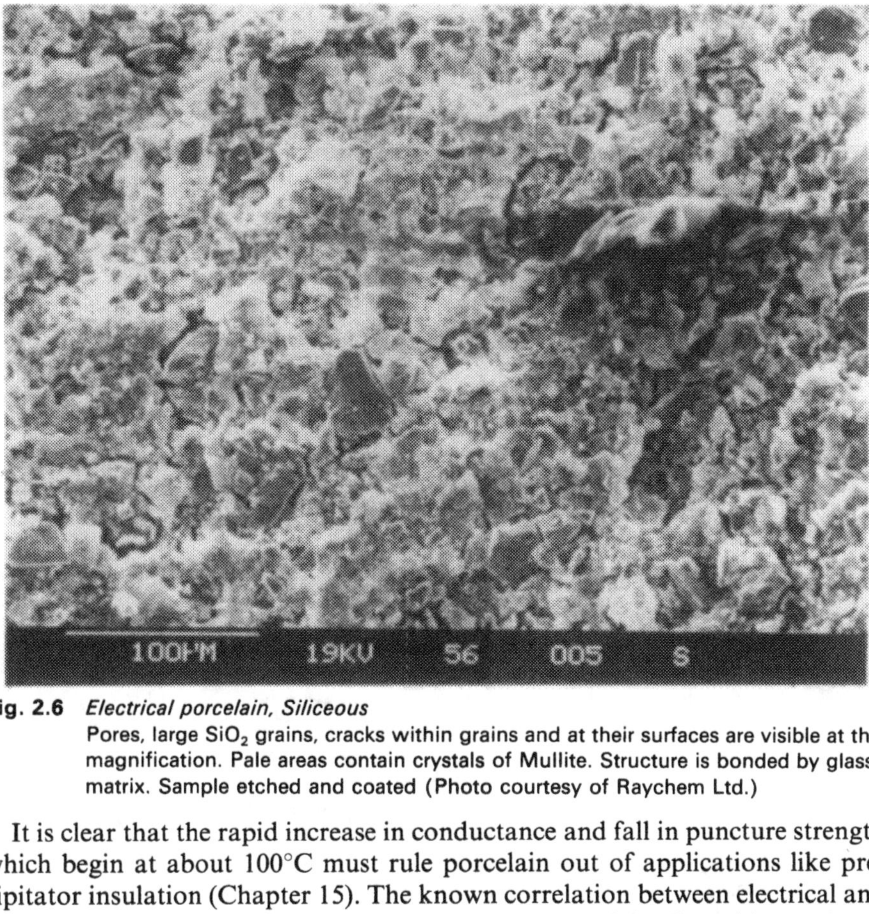

Fig. 2.6 *Electrical porcelain, Siliceous*
Pores, large SiO_2 grains, cracks within grains and at their surfaces are visible at this magnification. Pale areas contain crystals of Mullite. Structure is bonded by glassy matrix. Sample etched and coated (Photo courtesy of Raychem Ltd.)

It is clear that the rapid increase in conductance and fall in puncture strength which begin at about 100°C must rule porcelain out of applications like precipitator insulation (Chapter 15). The known correlation between electrical and mechanical strengths and the fact that puncture strength is affected by mechanical stress[27], when considered together, suggest that the reliability of porcelain insulators which are both mechanically and electrically highly stressed may be less than expected, from experience of less stringent installations.

2.3 Properties of insulator glass

2.3.1 The glassy state

The physical nature of a glass is difficult to define. P. M. Hogg's pioneering paper on glass insulators, written[11] in 1939, quotes an early writer's description of glass as a 'concrete juice': a vivid statement not easily improved upon.

Insulator glasses are based on silica, although glassy forms exist of many materials both organic and inorganic and ranging from toffee to metal[28].

Glasses behave in many ways like undercooled liquids and have a liquid's lack of long-range order in the constituent atoms. Depending on the rate at which a molten material is cooled, molecules are trapped in different degrees of disorder. The resulting glasses have different densities: fast cooling gives lower density than slow. Annealing, by allowing the temperature to fall very slowly indeed, produces a state of maximum density little lower than that of the crystalline form. Zachariasen[29] nicely illustrates the broad difference between a crystalline and glassy form of the same two-element material (Fig. 2.7); the introduction of a third species of atom, for example sodium into a silica glass, permits the formation of an open network by providing positively charged ions to balance the unattached oxygen ions (Fig. 2.8).

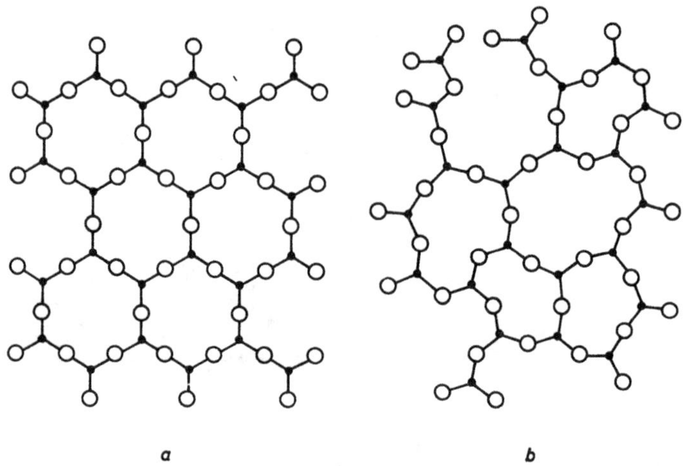

Fig. 2.7 *Zachariasen's models of crystalline and glassy forms*
Two-dimensional analogues of substance R_2O_3
a Crystalline
b glassy
● radical R
○ oxygen

Annealed glass is not a mechanically strong material because its surface contains microscopic cracks which are able to propagate freely through the volume, once they attain critical size, since there is no granular structure, as in porcelain, to stop them. Insulators made of annealed glass are therefore confined to classes of low mechanical rating. Where there is thought to be a risk of thermal cracking, borosilicate glasses of low expansibility are sometimes used. By far the greatest number of glass insulators now in service, however, are made from thermally toughened glass (Chapter 4) in which surface crack formation is suppressed by the production of a highly compressed skin.

Glass is toughened by heating it above its glass transition temperature and then rapidly cooling its surfaces, which first contract, leaving the interior in compression. Since, however, they are then 'frozen' into shape whereas the interior is still able to contract, the final state is of internal tension and surface compression (Fig. 2.9). Simple shapes like sheets have well defined final patterns of stress with a parabolic form of distribution. Complex shapes, like those of most insulator shells, do not have simple stress distributions because of the technical difficulty of cooling convoluted surfaces surrounding volumes of variable thickness. Great skill is therefore needed to produce reliable toughened insulator shells, and defects like flaking, where crescents of glass detach themselves from the periphery, are not unknown.

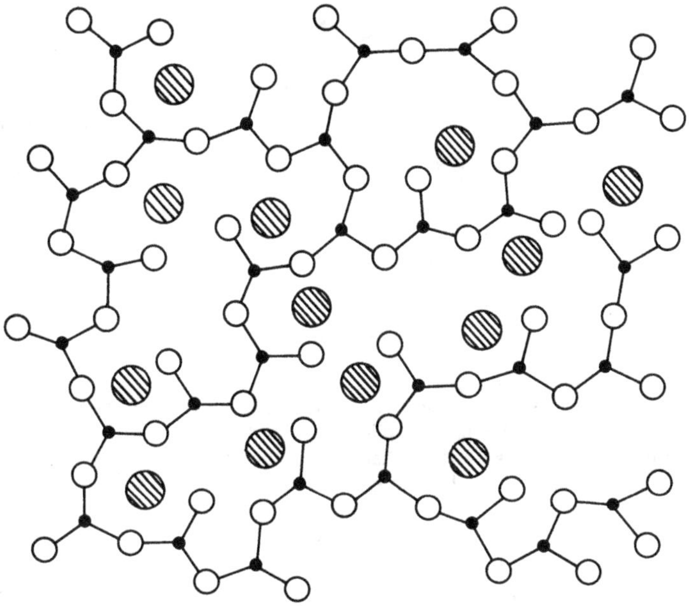

Fig. 2.8 *Glassy structure modified by alkali*
Addition of alkaline oxide, AO, opens up network, much reducing melting point, viscosity at given temperature etc.
● alkali ion A
○ oxygen

Statements about the mechanical or electrical properties of toughened glass must therefore be received with some caution since the physical states of both the compressed surfaces and stretched interiors tend to be variable and are, in practice, very difficult to establish, optically or otherwise.

2.3.2 Mechanical properties of insulator glass

The soda-lime silicate glasses, containing potassium, barium and aluminium as 'improvers', which are commonly used, toughened, for insulators have theoretical strengths about 7 GPa which are reduced in annealed specimens, by surface flaws, to values between 30 and 90 MPa. Toughening raises these to 100–120 MPa, high when compared with those from porcelain of 20–70 MPa for unglazed and 30–80 MPa for glazed. These are values of tensile strength: in compression both toughened glass and porcelains are at least five, and sometimes as much as fifteen, times stronger. Young's modulus is somewhat lower for insulator glass than for the best porcelains, varying between 60 and 70 GPa for the glass as against 50–120 GPa for the porcelains.

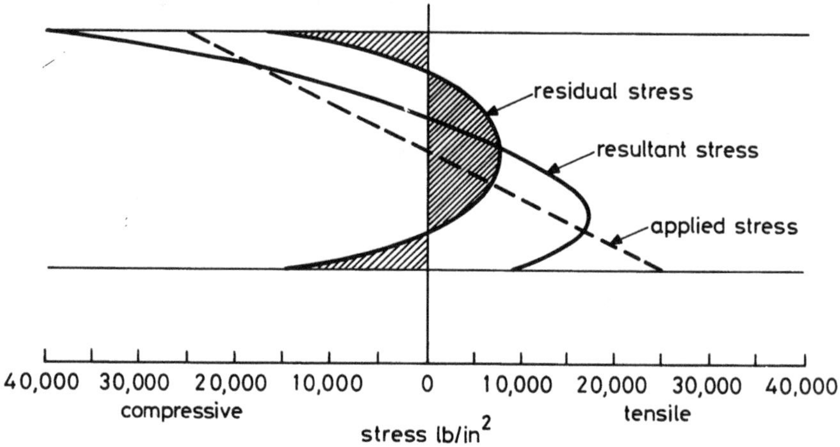

Fig. 2.9 *Stresses in toughened-glass plate (Reference 58)*
The distributions across a transverse loaded plate of toughened glass, of residual stress 'frozen' into the glass, applied stress and resultant stress. Insulator shells have more complex stress patterns than plates: they are much weaker when containing bubbles, inclusions and knit lines which are uncommon in plates

The thermal expansibility of insulator glass, at $8.5/10^6$ °K, is close to that of the cast irons, either malleable or ductile, from which caps and flanges are generally made[30]. The metal's expansibility is only some 35% above that of the glass, whereas for porcelains the ratio may be as high as 2·8:1. Well matched expansibility is a useful aid towards good performance under thermal cycling (Table 2·2).

The outstanding and crucial mechanical property of toughened glass is that it is in dynamic equilibrium, not static. Considerable strain energy is stored, in a glass insulator, as compression in the skin and tension in the body. If this is released, as it most commonly is by impact from a missile or gross erosion of a skirt, the whole piece flies into roughly cubical fragments. Disc insulators do not lose their mechanical strength because of this burst, since the fragments of the insulator head are retained within the metal cap. However, the debris from the rest of the disc may cause damage to property lying close to the tower – in the UK this effect has limited the application of glass in car parks and public places – while the total loss of creepage path from the insulator leads to enhanced leakage current over the remaining discs and thus to potential erosive runaway, if broken units are not promptly replaced.

Typical examples (Fig. 2.10) of erosion which has arisen in desert and marine areas include channelling of the surface and skirts, especially the inner one. The effect does depend on leakage current since units which are by-passed electrically in a string remain unaffected even when others of that string shatter because of erosion.

Fig. 2.10 *Glass erosion: three months' exposure to severe salt pollution*

Loss of the surface skin by ablative or 'sand-blast' erosion is sometimes advanced as a possible cause of shattering of glass insulators in desert areas. Against this speculative disadvantage must be set the fact that glass has a lower emissivity than many glazed porcelains, and will therefore lose less heat by radiation to the night sky. Flashover caused by dew is an important source of line outages in arid regions: retained heat inhibits dew.

2.3.3 Electrical properties of insulator glass

Since the electrical properties of porcelain are dominated by those of the glassy component, which has some common features with alkali–lime–silica insulator glass, similarities would be expected and are seen (Table 2.3). Both materials have relative permittivities of 5–7, both rapidly lose resistivity and puncture strength with increasing temperature and both show rising loss tangents with rising temperature and falling frequency. For glass, the increase in puncture strength, attributable to the absence of internal discontinuities which can act as electric-stress raisers, is, however, considerable.

The improved mechanical strength of glass when toughened, in comparison with porcelain, allows thinner shells to be used, on disc insulators. This leads to an increase in working electrical stress as high as 40%. Under AC stress the dielectric losses are correspondingly increased, which raises the question of thermal runaway in insulators whose capacitance has been deliberately enhanced by the use of unduly thin sections. Under DC energisation the high stress and high mobility of sodium ions lead to migration which may be in sufficient quantity to weaken the head mechanically (Chapter 15.4).

Even with the reduced thickness of dielectric, however, the electric strength of insulator glass is such that impulse puncture is virtually unknown, in disc insulator heads, although it may occur externally at the tips of discharges. Since puncture anywhere on the disc allows it to shatter under its own stored energy, it is not possible for hidden defects within the head to exist, as it is with porcelain.

The rise of electrical conductivity with temperature is so rapid, in soda glass especially, that thermal puncture under heavy pollution sometimes occurs, especially where electrical discharges are blanketed beneath layers of dirt. This may well contribute to the very severe erosion sometimes seen on glass insulators; it is, however, established that arc-damage, where the discharge is forced against a surface, is greater for materials like glass, of low melting point and thermal conductivity, than for refractories which are both less fusible and more conductive[31].

The glazes which are used on porcelain insulators are very much less liable to erosion than insulator glass. Here, the much smaller percentage of sodium in the structure may well give benefit: common sodium-rich glasses are rapidly attacked by the hot steam which evidently accompanies surface electrical discharges, when these occur in parts of wet contaminant layers[32](Fig. 2.11).

2.4 Properties of resin-bonded glass fibre (RBGF)

2.4.1 Fibrous composites: general

The strengthening of materials by the incorporation of fibres is now a science in its own right. Only the basic aspects which are relevant to insulator applications can be covered in this book.

In the principal applications of fibrous composites to insulators, respectively polymeric longrods and live-line accessories, effectively all the mechanical loads are carried by the fibres; the function of the resin matrix is to protect the fibres and to transfer load from those which are broken. For these applications, but not for others, the matrix can be much less strong and stiff than the fibres. Common matrix resins are polyesters and epoxies, both of which are viscoelastic to some degree. Most matrix resins change their properties with time and many are also subject to temperature effects.

They are required to protect the glass fibres against invasions which could cause microcracks – in particular by water, which is also highly unwelcome for electrical reasons – and to cushion lateral contacts between fibres.

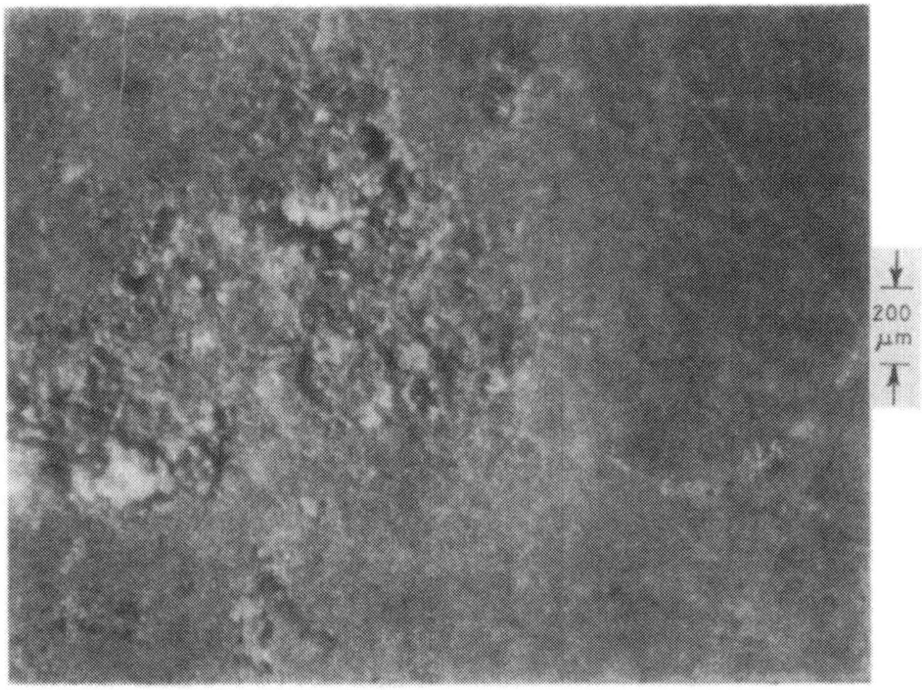

Fig. 2.11 *Deterioration of ceramics*
The glazed insulator-porcelain has undergone 14 hours of discharge attack in TERT procedure. Apart from arc-root melting there is negligible erosion. Soda-glass samples are severely attacked in much shorter exposure

Of equal importance to the properties of the matrix resin are those of the priming material, which is used as a coating on the fibres to promote chemical bonding between resin and glass and to exclude water vapour from the glass. Organosilanes, molecules which key tightly to the glass at one end and to the resin at the other, are generally used, each one specific to the particular glass and resin. These primers have been found to play a decisive role in the maintenance of high electrical resistance following immersion in water[33].

The glass fibres themselves fall into classes depending on their chemical composition (Table 2.4) and intended use. Because they are made by drawing from the melt and are protected immediately on emergence from the molten state by a layer of starch-like sizing material, their surfaces have few defects and the defects tend to lie axially rather than radially, Tensile strengths not much lower than the theoretical ultimate, of some 7 GPa for alkali glasses, are realised in virgin glass fibres. Otto[34] measured tensile strengths up to 3 GPa for E-glass, the type most used for electrical insulators, or about two decades higher than the values obtained for ordinary glass. Recent values are even higher.

Table 2.4 *Some glasses used as fibrous reinforcement*

Glass	Density (g/cm^3)	Tensile strength (GPa)	Tensile modulus (GPa)
S	2·49	4·0	85
E*	2·54	3·45	72
D	2·16	2·4	52
Quartz	2·2	1·7	72

* ECR glass, a corrosion-resistant variation on E-glass, has about the same mechanical properties as E. It contains a few per cent Zn but much less B than the unmodified glass.

It is, however, the properties of imperfect and contaminated RBGF which are of practical concern. There are more than one million fibres in a typical pultruded rod of 20 mm diameter, each about 10 μm diameter, giving a very large area of bonded glass, about 30 m^2/m length. Since the surface resistivity of most glasses falls rapidly with increasing humidity[43], invasion of a RBGF sample by water is likely to produce a large fall in electrical resistance. For similar reasons, a sample exposed to surface contamination and electric stress is liable to serious damage by leakage currents. An intact surface, either of glass and resin for a centreless-ground piece or of resin alone, when the sample is left 'resin-rich', will survive tracking tests for a while without distress but will fail rapidly once the surface is undermined. The leakage increases to levels which will cause tracks, burns and even incandescence in exposed fibres themselves (Fig. 2.12).

Imperfections are difficult to avoid in RBGF, again because of the large specific area of glass surface and the necessity to wet, with matrix resin, the whole of this surface. Axial porosity is often found (Fig. 2.13), which leads to wicking of contaminants and promotes internal discharges under high electric stress. The long, narrow pores which occur in RBGF are dangerous both when air-filled and water-filled. The air-filled void, being surrounded by material of much higher permittivity, suffers enhanced electrical stress when the sample is subjected to the high intensities which occur at terminals and below dry bands. When discharges occur within such voids, chemical decomposition of the surrounding polymer occurs, the products of which, according to Reynders[36], may include oxalic acid, a known promoter of mechanical failure in glass fibres.

If the pore has been invaded by water it becomes a needle electrode, at the tip of which the electric-field intensity may be very high: Vibrans[44] gives a multiplication factor, above the impressed intensity, given by the quotient of needle length by tip radius (Fig. 2.14).

These basic considerations, of stress enhancement either within an air-filled void or at its tip, for a water-filled one, are insufficient to explain the observed cases of destructive propagation within RBGF. The true case is much more complex. Recent work by the author and his colleagues has shown, by means of artificial faults, that several different processes may occur[45].

Since some flaws exist in the fibres, their strengths are both statistical and time-dependent. It follows, therefore, that the properties of RBGF will deteriorate with time both for this reason and because of the time dependence of the properties of the resin. For some applications, as in aircraft, extreme values of strength/weight ratio in structural materials are required, and lifetime is limited, by other considerations, to a few years. For insulators, however, longer lives are essential while the cost penalty of weight is much less. Insulators based on RBGF have therefore fallen into disfavour, where mechanical failures have occurred after quite short periods of service. Some of these failures have not been caused by the steady decline in strength, as mentioned above, but by a catastrophic fracture process, not yet fully elucidated but apparently arising from localised invasions by ions. These occur at cracks or other breaches in the protective housing: they affect fibres, matrix, or both[35–38].

Fig. 2.12 ***RBGF attacked by pollution and discharges***
Less than 1 hour under TERT

2.4.2 Unidirectional fibre materials

In most polymeric insulators the core uses unidirectional fibres, parallel to the axis, constituting some 70% by weight of the composite and made by the process of pultrusion. Large numbers of continuous fibres – a typical 20 mm rod will contain more than a million – are fed under tension through liquid resin which is thermally cured in a die: the process combines pulling with extrusion. Pultruded tubes which contain several layers of axial filaments, sandwiched between helically wound continuous 'barrel hoops', are used as live-line insulators[39]. These are generally based on cylindrical mandrels of closed-cell foam allowing stiff structures to be achieved which weigh much less than solid rods of equal rigidity.

Remarkable tensile strengths are achieved in these cases (Table 2.5).

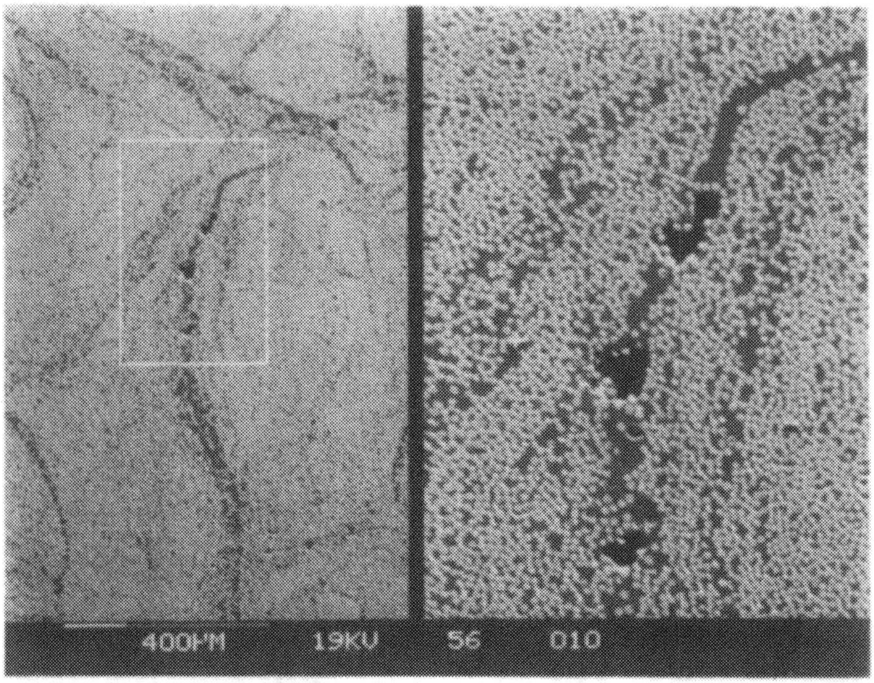

Fig. 2.13 *Axial porosity in RBGF*

2.4.3 Mechanical properties of RBGF

The structure of pultruded RBGF confers anisotropic properties both mechanical and electrical, and introduces some major differences from those of porcelain and toughened glass. In particular, the great tensile strength is not accompanied by equivalent compressive strength, whereas in ceramics the compressive strength is always greater and sometimes much greater (for some ceramics as much as 15 times).

Depending on the fibre content the axial Young's modulus ranges downwards from 90 GPa, which, together with the large tensile stresses commonly employed, results in strains far exceeding those in general engineering materials. A metre-long insulator core will typically extend by 10 mm or more before failing, thus placing heavy demands on the bond with any housing around it. Again because of the anisotropy, RBGF cores, and especially tubes, are easily damaged by crushing, when gripped under radial pressure or mounted in metal terminals.

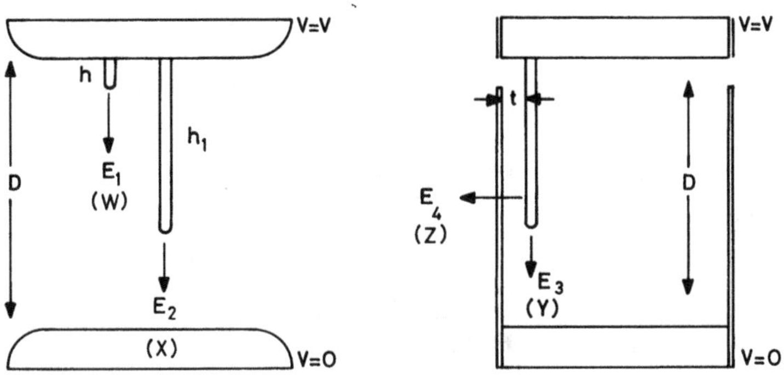

Fig. 2.14 *Stress multiplication by needle inclusion*

			Field intensity
(W)	Short projection into uniform field of intensity V/D: spherical tip, radius r	(axial)	$E_1 = (V/D)(h/r)$
(X)	Long projection: hyperboloidal tip, radius r	(axial)	$E_2 = 2V/r \ln (4[D - h_1]/r)$
(Y)	Long projection inside conductive cylinder, spaced by t, hyperboloidal tip, radius r	(axial)	$E_3 =$ approx. $2V/r \ln (4t/r)$
(Z)	As (Y) but between projection and cylindrical wall	(radial)	$E_4 =$ approx. $V/r(2t - r) \ln ([2t - r]/r)$

Torsional stresses, such as arise when one end of a core is fixed while the other is rotated, a common practice when power lines are strung, are especially undesirable. The internal splitting which may arise by failure of adhesive bonding between fibres becomes an insidious water trap.

The time dependence of tensile strength in RBGF arises from changes in both matrix and fibres. The loss of strength varies with the nature of the matrix polymer, the quality of the glass and of the primer, the thermal history and the loading history. Lifshitz and Rotem[40] showed that a polyester matrix having high creep rate and low creep compliance caused a faster loss of strength than did an epoxy of superior creep properties. The effect was large: the polyester lost 29% strength in two months' trial against the epoxy's 12%. The fibres will fail progressively with time both because they contain flaws which grow to critical size under load and because some degree of water-vapour penetration will occur. Water has so low a surface energy, in comparison with glass, that its presence in a flaw accelerates the propagation to failure. Poor primer, the main function of which is to modify the surface chemistry of the glass, or poor application increases the risk of weakening by water.

Elevated temperature generally accelerates the loss of strength, although there may be occasional improvements caused by slow curing of some matrix polymers, but a combination of high temperature and high humidity is almost always seriously damaging to strength. The mechanical-loading history acts in two ways. The number of flaws in the glass which grow to failure rises with load. The water permeability of the matrix generally rises with strain[41,42].

Table 2.5 *Mechanical properties of some RBGF materials*

Material	Make-up (%, v./v.)	Tensile * strength (MPa)	Tensile * modulus (GPa)	Expansibility ($\times 10^{-6}$/°C)
Rod: E-glass in high-grade aligned pultrusion	Fibre-65	1600	52	7·5 = 20·0 ⊥
Rod: as above; less glass, different resin	Fibre-45	950	35	7·5 = 29·0 ⊥
Sheet: E-glass, unidirectional	Fibre-65	> 700 = 30 ⊥	48	4·0 = 21·0 ⊥
Sheet: E-glass, 50/50 bidirectional	Fibre-65	480	30	11·0 = 36·0 ⊥
Tube: E-glass in wall, 50/50 axial/peripheral	Fibre-65	490	30	

* Transverse compressive strengths may be no more than 15–25% of tensile
Notes: Value parallel to rod axis or sheet plane =
Value normal to rod axis or sheet plane ⊥

Analysis of the strength decay is complex even for simple tensile loads, but for compression and cantilever conditions it is intractable. Cyclic loads and impacts are common features of insulator duty: these too may have large degrading effects on strength[17].

The insulator designer must do his best to assess the rate at which his core will weaken, from short-time tests by extrapolation, from theory and from quality control. His task is a difficult one, however, and it is not surprising that large factors of safety, in cross-sectional areas of cores, are usually applied. Practical testing philosophies are discussed in Section 12.6.

2.4.4 Electrical properties of RBGF

As long as a fibre-reinforced sample is in dry and undamaged condition its electrical properties are what would be expected from its make-up. Puncture strengths – not usually practically relevant – are of the order of 10 kV/mm; permittivities depend, in unidirectional fibrous materials, on orientation, being dominated by the glass, of relative permittivity about 7, along the axis but reduced by the multiple layers of polymer across it; loss tangent is an average of the polymer and glass values, which, depending on glass fraction, is usually about 10^{-2} (Table 2.6).

When a dry 'spike' was introduced, in the form of a metallic needle, either within or on the surface of RBGF samples of different qualities, and a large voltage applied to it, discharge activity began, which was readily detectable in the external circuit, or observable as a glow when transparent covers were used. Propagation did not occur, however, in exposures of many weeks' duration. Substitution of a water-filled hypodermic needle for the dry one (Fig. 14.9) caused some samples to fail in minutes, while others survived for hundreds of hours. Dissection showed that all the samples had suffered damage, commonly as axial burns, linked by sideways punctures, presumably between separate paths of weakness, or voids.

This type of water-induced failure was found not to be correlated with the surface vulnerability of the samples, as measured by TERT (tracking and erosion resistance tests) (see Table 2.10). Various fillers, which had been introduced with the object of improving resistance to surface leakage current, acted adversely in the presence of the advancing internal tracks.

When a second type of artificial fault was used, a capillary-wetted thread to simulate an end-to-end leakage path, again all the samples failed but with large variations in survival time. Similar evidence was found to that in the needle experiments, of water permeation followed by successive radial jumps and axial tracks. In this case secondary invasions could be seen to meet the advancing primary ones, but from the opposite electrode (Fig. 2.15).

In none of these experiments with RBGF was external pollution applied, as would universally be the case for real polymeric insulators. Experiments with polymeric housings on ceramic substrates showed, however, that great accelerations of destructive propagation do occur when the advancing internal track is capacitively coupled to an external 'electrode', in the form of a layer of conductive solution or deposit (Fig. 2.16).

2.5 Properties of polymers and polymer concretes

2.5.1 Applications and functions

The two principal classes are those where mechanical support and electrical insulation are combined, and those where the polymer serves only to to control surface leakage or protect internal parts. Most polymers are tens or hundreds of times more expensive than ceramics, weight for weight, and have low elastic moduli. Bulky, unreinforced polymers are therefore practically confined to low-voltage applications or to bushings, where a single part can combine the two functions of housing and stress grading, which call for two parts where

Table 2.6 *Electrical properties of some RBGF materials*

Material	Make-up (%, v./v.)	Puncture strength kV/mm, 50 Hz at (°C)	Loss tangent ($\times 10^{-3}$ at (°C))	Permittivity (air = 1, at (°C))
Rod: E-glass, high-grade pultrusion	Fibre-65	3·0 = (20) *	20·0 = (20)	5·3 = (20)
		14·0 ⊥ (20)	30·0 ⊥ (20)	3·7 ⊥ (20)
Sheet: E-glass 50/50 bidirectional	Fibre-65	2·4 = (90)		
		16·0 ⊥ (90)	30·0 ⊥ (20)	5·0 ⊥ (20)

* Axial puncture strength is variable, depending on porosity, humidity, history of piece

Notes: Values parallel to rod axis or sheet plane =

Values normal to rod axis or sheet plane ⊥

porcelain and foils are used conventionally. Cast epoxy, usually loaded with fillers, is used for bushings and polyester for the low-voltage types.

An exception is the Metapol design (Chapter 17) where polyolefines are used in combined mechanical and electrical function.

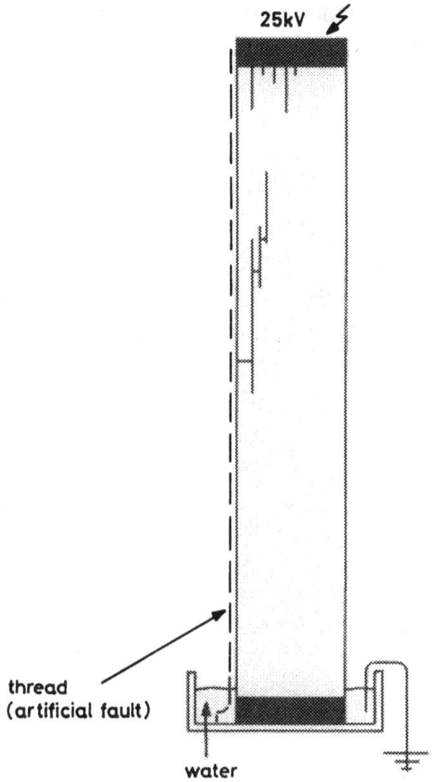

Fig. 2.15 *Water Permeation into RBGF under electric stress*
Failure occurred by sequential radial invasions and axial propagation along faults or voids

By far the greatest use of polymers is in housings, which surround mechanically strong fibrous composite cores, where the requirement is to control external leakage while protecting the contents against invasion by water and damage from leakage currents and discharges.

The polymer concretes, of which the main types are Polysil and NIM, are stiffer and stronger than most polymers. They are designed to perform both functions; however, present indications are that their electrical properties, in pollution, may generally be too weak for either to operate successfully without some form of surface protection.

In the recently developed Hybrid insulator, polymers are used solely to control the leakage current and with little if any protective requirement, since the substrate is not a vulnerable RBGF but a highly resistant ceramic.

2.5.2 Polymers for housings

The main functions of the polymers which are used to house RBGF cores are to protect the interior fibrous composite against water and to provide the working profile of the insulator. The profile must confer the requisite creepage path length, give minimum aerodynamic catch of airborne pollutants, and shed water both condensed and precipitated.

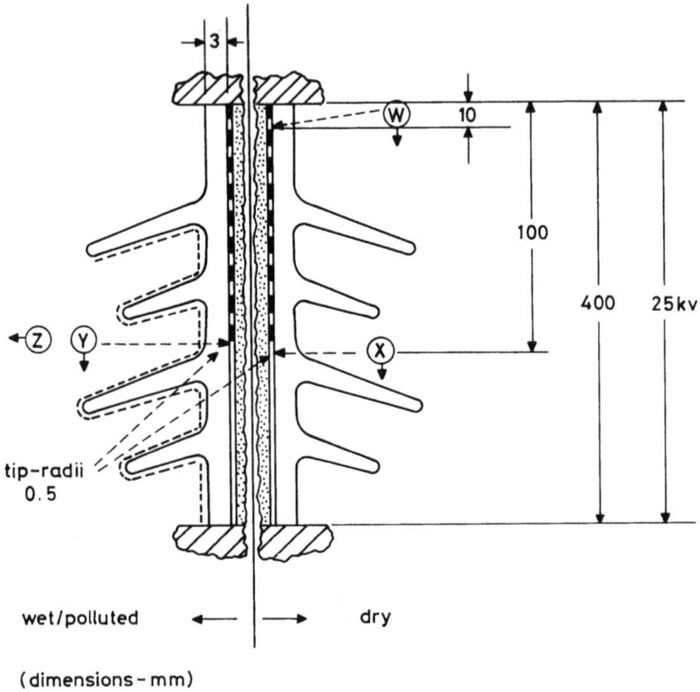

Fig. 2.16 *Internal propagation of damage, polluted and dry*
Dimensions in mm
In 'polluted' case (left), tip stresses *Y* and *Z* depend on local potential difference
Continuous pollution layer gives voltage gradient. Interrupted (shown) gives worst case
Typical practical values for given dimensions are: *Y* (axial) 31·47; *Z* (radial) 3·79 (kV/mm)
In 'dry' case (right) tip stresses *W* and *X* increase with penetration but are smaller
Typical values are: *W* (axial) 1·25; *X* (axial) 12·85 (kV/mm)
(Note: stresses *W* or *Z* are insufficient to break *air* down; *X* and *Y* approach and exceed, respectively, puncture strengths of *polymers*.)

In so doing the polymer must not track, must withstand weathering without undue loss of mechanical or electrical properties, must erode at an acceptable rate and in a tolerable manner, and must resist mechanical loads, both static and cyclic, as well as temperatures from -40 to as high as $+120°C$.

Tracking is the production of a resistive path by degradation of the polymer, usually because of deposition of carbon; erosion is the loss of surface material which arises mainly from electrical discharges or from electrochemical reactions in the layer of contaminant. Tracking and erosion increase with density of leakage current, and the severity of both can be limited by reducing the surface electric stress; in effect, there is no such thing as a 'non-track' polymer, but tracking can be minimised by avoiding known carbon generators like benzene rings or counteracted by arranging for liberated carbon to be scavenged by chemical means. One such is the use of active fillers, like hydrated alumina in conjunction with other catalysts, to oxidise the carbon into carbon monoxide and water vapour. Housing material, both polymer and filler, is lost in this process, but a fresh and non-conductive surface is exposed[46].

Table 2.7 *Predicted lifetimes of housing polymers: outdoor tests*

Material	Predicted Lifetime (yrs.)	Observed Damage
PTFE	20	Slight erosion
Silicone elastomer, best	25	Tracks and punctures
Silicone elastomer, worst	2·6	Tracks
Polyolefine, best	5	Puncture
Polyolefine, worst	1	Tracks

Note: These data were obtained from 25 kV samples under accelerated wetting.

Electrochemical erosion is caused by the discharge of the ions, which carry the surface leakage current, when they meet either the metal terminals or ends of carbon tracks, Aggressive products of this electrolysis and later hydration include NaOH, which is known[47] to etch some materials, especially filled epoxies, causing holes or channels. Whereas diffuse erosion is usually tolerable on the surface of a housing and may even give some degree of self-cleaning[48], localised or concentrated erosion is usually [49] fatal since it allows invasion of the interior by water and contaminants.

The criteria for a good housing polymer are seen to be complex and contradictory. A non-wetting and non-stick surface will reduce leakage current in pollution but give poor bonding to the protected substrate: fluorocarbons like PTFE/Teflon are an example. A powerfully adhesive resin like an epoxide will give excellent bond to the substrate, but, in usual filled forms, will adhere strongly to surface dirt and water and consequently will suffer rapid surface degradation: epoxies also tend to embrittlement at low temperatures. Some apparently suitable families, like the polyolefines and copolymers, need to be protected against oxidation, during processing as well as in service. This calls for the incorporation of antioxidants and UV absorbers which, in turn, degrade the primary properties [50].

Table 2.8 *Deterioration of housing polymers in 20 months' marine pollution*

Material	Roughness (μm)		Water contact angle (rad)		Surface resistance: RH = 90%, (GΩ)		Salt-fog flashover (kV)		Leakage currents	
	New	Aged	New	Aged and cleaned	New	Aged and cleaned	New	Aged and cleaned	Weeks to first 500 mA surge	150 mA surges count rate: porcelain = 1
Porcelain	0·13	0·13	0·98	1·15	400	0·4	7·6	7·8	44	1
PTFE		1·35		1·65				7·1	49	2
Silicone elastomer			1·8	1·75	6000	400	10·9	10·8	> 86	0·2
Polyester DMC; filled with ALTH	0·15	1·3	1·3	0	2000	0·002	8·0	5·4	17	4
Epoxy, alicyclic; HHPA cure, silica filled	1·05	2·5	1·7	0	10000	0·01	10·9	6·9	19	13
Epoxy, as above, but ALTH filled	0·81	1·8	1·55	1·1	2000	3	8·3	7·3	8	14
Epoxy, bisphenol; HHPA cure, silica filled	1·0	1·9	1·5	1·25	10000	0·05	8·7	7·0	15	6
Epoxy, as above; amine cure, silica filled	1·05	Failed	1·55	Failed	3000	Failed			5	14

Since all housings have to be manufactured and fitted in some way to cores and metalwork, facility of shaping and compatibility with other, dissimilar, materials are seen to be prime requirements. With all these constraints it is not surprising that simple polymers are never used for housing but always mixtures, sometimes very complicated ones, of several principal polymers, fillers, ancillaries to confer heat retractability, pigments and the above mentioned antioxidants and UV absorbers. It is sometimes overlooked that the actual processes of curing and forming introduce further substances, like the residues of polymerisation and catalysis, and that benign additives like alumina are sometimes contaminated with undesirable ionic salts. Practical housings can therefore not be simply characterised by their main component: it is meaningless to talk about 'silicone rubber insulators' when the variation of life expectancy between best and worst is far greater than between one silicone and one polyolefine (Tables 2.7 and 2.8).

In Table 2.9 some properties are given of housing materials, most of which are currently in use on practical insulators: data from simple polymers are included for comparison.

2.5.3 Evaluation of polymers as housing materials

The evaluation of polymers which are candidates for use as housing materials comprises two stages, respectively the weeding-out of intrinsically objectionable classes, such as those which track readily or take up large amounts of water, and the attempt to predict the behaviour of a material from tests on virgin substances and on those which have been artificially aged.

'Accelerated testing' is presented in the literature as a real concept, especially by manufacturers, and statements are made such as 'a life of thirty years may be expected on the evidence of this test'. Predicted performances should, in fact, be treated with suspicion even for single polymeric pieces, while for complete assemblies, as of housing, core and terminations, the degree of synergy between the different mechanisms of aging is such that exposure of complete insulators to severe pollution under some degree of electrical overstress is, at present, the sole reliable guide. Even here the sensitivity of polymeric surfaces to the nature of the contaminant, as well as to its electrical conductivity, may mislead; some polymers are highly vulnerable, for example, to adhesive layers of low electrical conductivity, such as cement dust, especially where these interfere with the designed scavenging processes based on active fillers like alumina trihydrate.

The principal tests which are in current use for the evaluation of polymeric materials, in terms of liability to track or erode or both, are listed in Table 2.10. Two, ASTM D2303 and IEC 587, contaminate the sample by flowing electrolyte. Both use ammonium chloride to avoid build-up of non-volatile salts and both adopt the same resistivity. The ASTM D2132 is a dust-and-fog test, using a synthetic dust containing 3% NaCl which is moistened by a fog of water having low conductivity (about that of most town waters). Drop tests, like IEC 112 and VDE 0303, supply drops, at 30 s intervals, of the same electrolyte

Table 2.9 *Properties of some housing polymers*

Material	Resistance to:				Processing	Remarks
	Tracking	Erosion	Chemical attacks	Weather		
Sintered PTFE: mouldings bonded with other fluoropolymers	Excellent	Good, but becomes wettable	Excellent	Excellent, but becomes roughened	Difficult	Bonding polymers may track. Low thermal expansibility. Tough.
EPDM filled with ALTH: mouldings bonded with EPDM	Good	Good	Good	Good	Needs vulcanisation	Heavy filler content, may lead to internal propagation, from defects or along interfaces.
Silicone elastomer filled with ALTH: hot cured mouldings	Excellent	Good	Excellent	Good	Difficult to bond mouldings	Depending on monomer and additives, may give good retention of non-wetting surface property.
Heat shrinkable silicone/ polyolefine: filled with ALTH and catalysts	Very good	Good	Good	Good	Easy	Purges surface carbon by catalysis. Needs good interfacial sealant.
Thermoset alicyclic epoxy filled with ALTH and silica	Good	Good	Vulnerable to NaOH	Poor	Easy	Housing subject to cracking under bending. Severe increases in surface roughness and wettability after weathering and discharge attack.

Silicone elastomer: 30% w/w silica filler	Unacceptable	Good	Good	Simple polymers. Both track readily, leaving heavy carbon char
Polyvinyl chloride, unfilled	Unacceptable	Fair	Poor	

Notes: EPDM = ethylene propylene diene monomer
ALTH = $Al_2O_33H_2O$: alumina trihydrate

specified for the inclined-plane tests, to the surface between close-spaced electrodes.

The comments, made by the authors of the tests themselves and incorporated in the Standards, are interesting. ASTM D2303 warns that 'The test conditions . . . do not reproduce all the conditions encountered in service. Therefore, inferences . . . should be made with caution'. ASTM D2132, on the other hand, claims 'This method is an accelerated test that is intended to simulate extremely severe outdoor contamination'. Of the other two tests, IEC 112 claims only 'good resolution between materials with relatively poor tracking resistance', and points to IEC 587 as appropriate to materials intended for use under severe ambient conditions.

Table 2.10 *Tests for evaluation of tracking and erosion rates*

Test	Main features
IEC 587	Inclined planar sample. NH_4Cl pollutant with wetting agent, sometimes also sugar, added. Energised at 1·5 – 6·0 kV
ASTM D2303	As IEC 587. Specifies procedures for initial tracking voltage, time-to-track, erosion rate
ASTM D2132	Dust fog test. Dust: SiO_2, clay, salt, paper pulp. Fog: about 1·0 mg/cm^2 min. 1·5 kV across 2·54 cm, sloping electrodes
IEC 112	Comparative tracking index test. Drops of NH_4Cl at 30 s intervals. Up to 0·6 kV
BS 3781	AS IEC 112 but wetting agent added and test voltage up to 1·4 kV
VDE 0303	As IEC 112 but uses wetting agent, Pt instead of brass electrodes. Two procedures specified
Markussen/ Paulsen	Dip-and-drain test. Samples are spokes on wheel rotating around inclined axis at 0·5 rev/min. Immersed: 20 s. Stress: 0·68 kV/cm. Pollutant NaCl solution
Ontario Hydro Ferris Wheel	Similar to Markussen/Paulsen but pollutant contains solids and is sprayed

An obvious objection to these tests arises from the severity of the electric field, imposed by high test voltages across small clearances. A real insulator in the field usually carries no more than 0.2–0.3 kV/cm along its surface creepage path, while the tests apply 1.2 kV/cm or more. Any surface-water repellence of the material under test will be unfairly over-ridden by these high intensities of field which cause spreading of drops otherwise contained by surface tension. Similarly the heavy pollution, which is forced upon all samples by the methods of contamination, discriminates against those materials which, because of their surface properties, would catch much less than others of the airborne dirt in real conditions of service.

Nevertheless, useful comparisons have been made by the use of some of these tests, particularly when they were performed in conjunction with artificial weathering. Thus Clabburn, Penneck and Swinmurn[50] made an initial selection from a range of polymers by inclined-plane testing, went on to use the same class of test to investigate the effects of additives like fillers and pigments, artificially aged some of the selection in a WeatherOmeter and tested again on aged samples. They also made thermographic analyses and observed the degradation in mechanical properties.

Other methods of evaluation use different cycling arrangements whereby samples are subjected to contamination, electric stress and some type of artificial aging or weathering, in sequence or combination. The object is to take account of the fact that, in service, insulators are indeed subject to large cyclic variations which lead, for example, to particularly severe conditions under drying because of the development of local electric and thermal 'hot spots'. Main features are shown in Table 2.10.

Table 2.11 *Some tests simulating real conditions*

Test	Reference	Main features
British Rail: Accelerated wetting		Natural weathering. Hourly water spray, 1·5 min duration. Service voltage. Time to failure noted
British rail: Iron oxide		Fog chamber. Sample precoated with artificial layer containing Fe_3O_4. Service voltage. Damage noted after 8 h clean fog
Detroit Edison: Cyclic test		Artificial weathering. Cyclic wetting. Service voltage. Time to failure noted
IEC: 1000 h fog test		Indirect fog of NaCl, 10 kg/m^3 salinity. Stressed at 34·6 mm of creepage per kV actual. 'Failure' criteria include overcurrent tripouts, tracking, erosion, shed puncture

Although simple mixtures of inert and ionic pollutant are sufficient for evaluating most housings, they are not for others, particularly those used in electric traction. Here there are special hazards from dusts generated by the vehicles themselves, and in particular from oxides of iron derived from the brakes. Tests for such housings are detailed in Table 2.11.

The final arbiter of whether or not a polymeric housing is acceptable is, of course, actual duty, outdoors under voltage. A useful intermediate step in evaluation is, however, short-term exposure to fog, preferably salt fog. Provided that proper sample geometries are chosen, and a power source capable of delivering several amperes at test voltage, weaknesses unforeseeable from the low-power small-size tests mentioned above are sometimes identifiable. Thus, a particular thermoset which gave excellent results in a standard TERT (tracking and erosion test) was found to ignite when subjected to high-power testing on a length of 1 m in salt fog of salinity 2.5 kg/m^3.

Outdoor tests, to evaluate polymeric insulators under exposure to electric stress, contamination and weathering, were made from an early date in England[51], both on dummy and real load-bearing insulators and later in continental Europe and the USA[52]. It cannot be said that good correlation has been found with results based on TERT and similar tests on housing materials. This has been because crude failures, by tracking or erosion of housings, have been overwhelmingly swamped, in frequency, by either electrical or mechanical collapse caused by interfacial invasions. These, almost always originating from terminals but sometimes, with multi-part housings, affecting joints and frontiers between skirts and bodies, have caused damage sufficiently severe to end the test, long before the bulk of housing has suffered significant deterioration.

Some types of housing material have, however, shown characteristic advantages and disadvantages. Details are given in Table 2.9, but extreme cases are the heavily filled thermoset epoxies, which generally suffer severe loss of surface properties causing a decline in performance, measured in flashover voltage per standard severity of pollution, as high as 40% from initial value in two to three years, and the fluorocarbons, especially PTFE (polytetrafluorethylene or Teflon). PTFE housings have, in many cases, lost some surface finish and become somewhat wettable, but otherwise have failed almost entirely by puncture to the core, or undermining of skirt/body junctions. The rubbery materials, such as the olefine-derived EPR or EPDM (ethylene propylene rubber; *e. p.* diene monomer) and the butyl rubbers have shown only fair maintenance of surface properties while some have tended to split under weathering. Elastomers having silicone rubbers as a principal constituent have often shown good maintenance of 'non-stick' properties, although mechanical weakness and bonding difficulties have plagued some.

In spite of the enormous amount of effort which has been expended on development and testing of housing polymers, it now seems clear that there is no 'magic' material and that a successful polymeric insulator can only follow solution of all the problems discussed below (Chap. 6).

2.5.4 Polymer concretes

Two distinct materials, both classed as polymer concretes but different in almost all respects, are currently under development as insulator dielectrics (Table 2.12).

Polysil, a development of the Electric Power Research Institute, is an acrylic polymer heavily loaded with sand. NIMS (acronym for new inorganic materials) are compacted hydraulic concretes containing polymer only as a rheological aid in processing.

The principal advantages claimed of Polysil over porcelain are its low-temperature, low-technology process of manufacture, its resistance to impact and its expansibility, which, being close to that of some metals, allows metal parts to be cast-in during manufacture. It is made by mixing dry aggregate with the acrylic-based monomer and hardener, vibrating and pouring into moulds and curing in times between ten minutes and 24 hours, depending on size and formulation.

Table 2.12 *Properties of polymer concretes*

Property	NIMS	Polysil
Flexural strength (MPa)	40–150	40–90*
Flexural modulus (GPa)	25–40	30*
Fracture energy (J/m^2)	50–1000	70–120*
Tensile strength (MPa)	30–100	35–60*
Density (g/cm^3)	2·3–2·5	2·2–2·4
Expansibility (($\times 10^{-6}$)/°K)	10	24
Compressive strength (MPa)	150–280	150–190
Thermal conductivity (W/m°K)	1	1
Electrical resistivity (Ω cm)	10^7–10^{12}⊕	$> 10^{12}$
Puncture strength (kV/mm)	8–13⊕	17

* These values deduced from data published by EPRI.

⊕ Resistivity is nonlinear: it and puncture strength depend strongly on dryness of sample, with NIMS.

The electrical performance of Polysil in pollution is not finally established. Apparently it survives light pollution and weathering without undue damage. Tests made in severe marine pollution have, however, led to rapid surface failure. Like other mineral-filled thermosets it offers an inevitable increase in both surface roughness and free energy with time. Rises in surface leakage current occur, with discharge erosion on an increasing scale, coupled with falls in flashover voltage.

The NIMS are not low-technology materials and their manufacture includes not only mixing but also treatment of the resulting dough, by moulding, extrusion or other high-pressure operations with heat-curing and drying. Unlike Polysil, NIMS cannot be used directly to form complete insulators because of their variation in volume resistivity with humidity. Both sealing of the concrete and use of an external sheath appear to be needed. However, the good resistance of NIMS to impact, and their ability to be simply bonded to metallic terminations, appear to offer prospects of their use in Hybrid insulators of high strength (Chap. 14).

The principal distinction to be drawn between NIMS and Polysil is in the polymer content – about 30% v./v. with Polysil but about 10% v./v. for NIMS – and consequently in the stability of the surfaces under weathering and discharge. Neither offers much promise of successful use without external protection, in pollution and especially marine type; neither really lives up to the early hopes of a material which could be cast, on site, to give a concrete structure combining both mechanical strength and electrical insulation.

Chapter 3

Manufacture of wet-process porcelain

3.1 Wet and other processes

Water is used as a carrier medium in practically all porcelain processes appropriate to insulators. Its function is to allow intimate blending of the main constituents, respectively ball clay, china clay, felspar and quartz. The significant difference arises in the stage at which the water is removed.

Where the intended shaping process is pressing, suitable grains or powders are prepared: either filter pressed cake is made from the slurry, to be granulated or dried and crushed, or the slurry is spray dried to give fine dust. Additives are incorporated, to promote flow under pressure, before the granules are fed into the cavity between metal faces in the press, or before the dust is compacted isostatically (i.e. in a compliant sock mould under all-round pressure)[54].

There is confusion in nomenclature: the above type of dry pressing is sometimes called 'hot pressing', but the same term is also used for a wet process, described below.

Only those shapes which can be removed from moulds or forming surfaces can be made by pressing, thus ruling out re-entrant or undercut profiles which are some of the most useful (Fig. 3.1). Machining after pressing must therefore be used for these, introducing a health hazard in the production of airborne dusts, when performed on dry body.

Of the wet processes (Fig. 3.2) casting, jolleying or jiggering and hot pressing are similarly restricted to removable profiles. Jolleying or jiggering uses a plaster mould to confer the shape, as does hot pressing. In the latter, a heated rotating die forms the other face.

Turning, either of cylinders or discs, allows advantage to be taken of the remarkable ability of clay-based bodies to accept almost any shape of profile and retain it during firing, albeit with some shrinkage and other limitations arising from the mechanical weakness of the material at high temperature.

Quite a wet body is needed for turning and hot pressing; in this book the convention is adopted of expressing water contents as parts per hundred by weight of the fully dry body.

3.2 Blending the raw materials

The principal raw materials are ball clay, china clay, felspar and quartz, roughly as 50% clays, 25% felspar and 25% quartz.

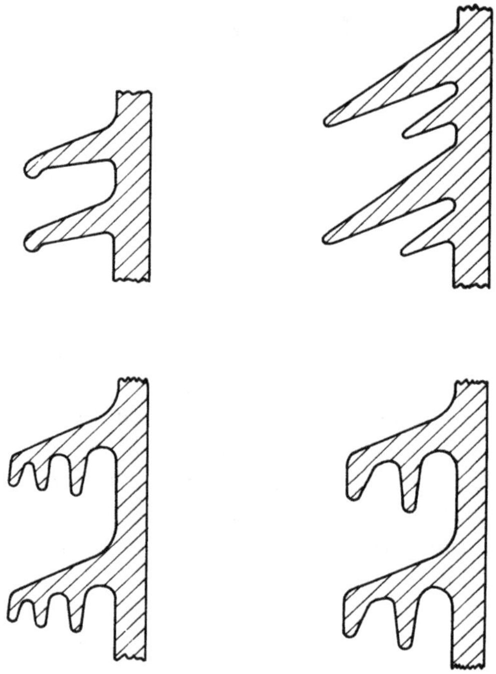

Fig. 3.1 *Common re-entrant profiles of insulators*
None of these common shapes can be made by pressing or moulding, in porcelain. Sections must be made and stuck-up, or shapes must be turned

Both ball clay and china clay are decomposition products of granitic rocks and both comprise mainly kaolinite, $Al_2O_3.2SiO_2.2H_2O$. Kaolinite is a layered silicate: even at molecular level its structure comprises planes of atoms, in a multi-deck sandwich assembly of oxygen, silicon, oxygen with hydroxyl, aluminium and finally hydroxyl. The remarkable rheology of clays arises both from this layered form and from the manner in which water is embodied into the material (Fig. 3.3). The content of very fine particles is higher in ball than in china clay: both are generally contaminated by compounds of calcium, iron, potassium, sodium and titanium, as well as by fine-grained quartz and carbon, which may occur in organic compounds or as carbonates.

Felspars are alkali–aluminium silicates, X $O.Al_2O_3.6SiO_2$, where X is either potassium or sodium. The choice of sodium or potassium, or their relative proportions in a mixed felspar, considerably affects the final electrical properties. The mineral is pre-crushed and delivered dry-bagged as powder of about 300 mesh, i.e. to pass a 45 μm aperture.

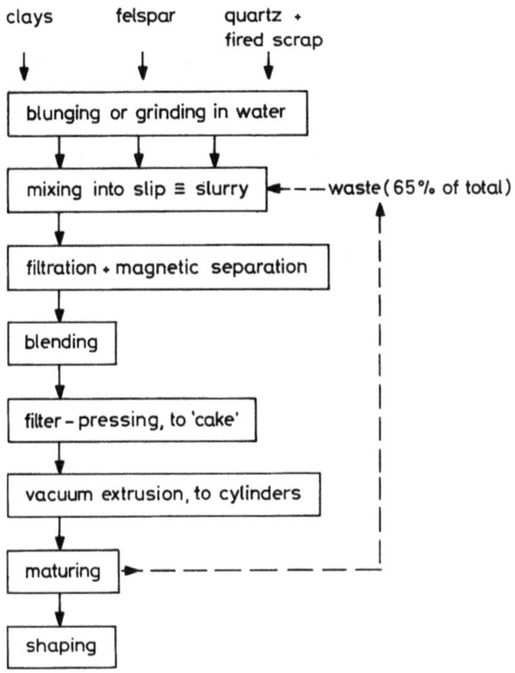

Fig. 3.2a *Manufacture of porcelain insulators: preparation of body*

Quartz, which constitutes the 'frame' on which the final product will fit and the crystalline form of which is crucial to the expansion characteristics of the body, is one of several polymorphic forms of silica, SiO_2. The mineral is milled in water, using flints as grinding balls, to a size distribution of 55% less than 10 μm.

Secondary but important semi-raw materials are scrap fired porcelain, ground to give 40% less than 10 μm and replacing quartz by up to 5%, and reworked unfired scrap body from residues of the extrusion and shaping processes. Some 55% of unfired body is recycled, in this way, with valuable randomising effects on the composition.

The clays require considerable energy to overcome their intrinsic self-adhesion and disperse them in water. An octagonal tank equipped with rotating blades and called a blunger is used for this purpose. The felspar is held in aqueous suspension as is the quartz: agitated containers (arks) are used to hold the suspensions to which traces of alkaline deflocculants, such as $CaCl_2$, are sometimes added, to delay settling of the solids.

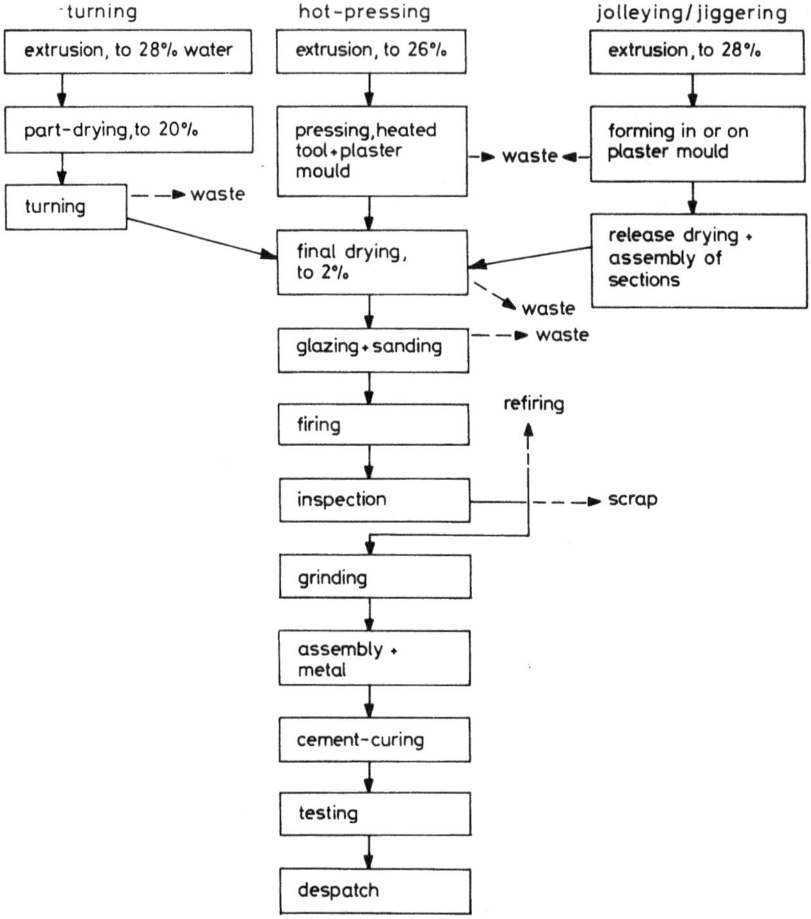

Fig. 3.2b *Manufacture of porcelain insulators: shaping, firing and finishing*
Note: A 'blunger' is an octagonal tank containing rotating paddles

Mixing is done on the basis of density measurements, from which dry weights of the components are calculable. Recycled unfired body, from the subsequent steps of the process and also blunged into a suspension, is added at this point. The blended liquids are passed over filtration lawns and magnetic separators, the latter to remove any ferromagnetic impurities.

3.3 Dehydration and forming

Much of the water is removed by filter pressing, sometimes with applied heat to reduce the viscosity of the water. The resulting cakes are extruded under reduced pressure, to remove entrapped air, by which time the water content is some 30%.

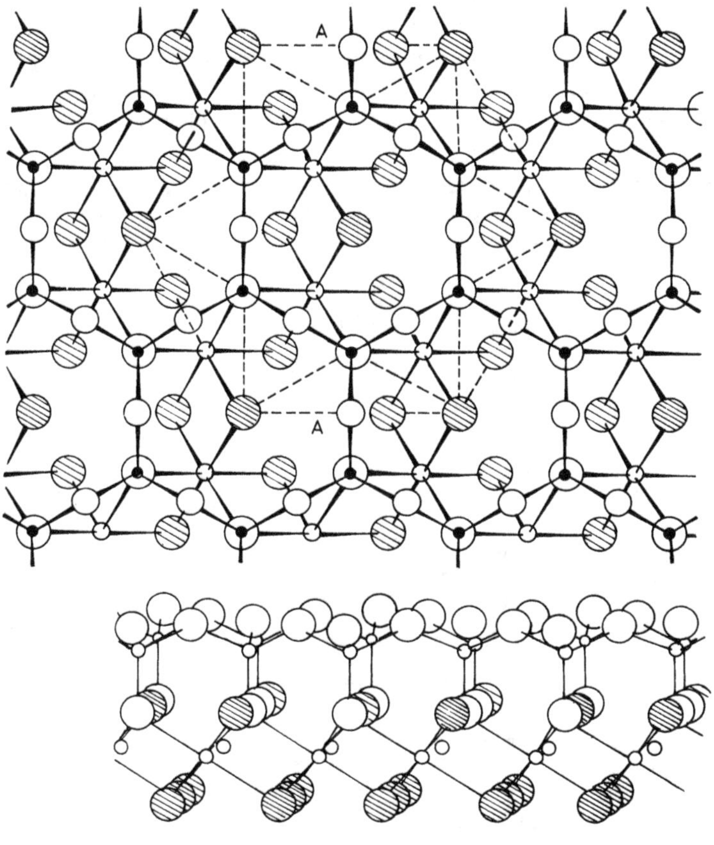

Fig. 3.3 *Structure of a clay – Kaolinite (after Reference 57)*
Plan and perspective are shown of one layer out of many, between A, A: distance is about 1 nm.
In descending order of levels are: O, Si, O and OH, Al, OH.

Storage, to mature the body and aid in uniformity and workability, is general practice.

The first step in forming is a second extrusion (pugging) to yield either solid or hollow cylinders, depending on the final requirement. These are cut to length – the body now has the consistency of cheese – and further water is removed. If turning is intended the water content is taken down to 20%; if hot-pressing, to 26%. A combination of circulating warm air and direct resistive heating of the body, by passage of current between end electrodes, is generally used for this highly critical stage. Too slow drying is costly in process delay; too fast is liable to produce cracks and flaws which, being internal and thus undetectable, are likely to cause unforeseen failures of product in the kiln.

Turning to shape is performed either in vertical or horizontal lathes. Good judgment is needed to avoid internal damage by torsion. The body, although soft in texture, is highly abrasive and calls for the use of carbide-tipped tools, or at least special steels, if unacceptable loss of profile-precision is to be avoided.

Hot pressing uses thick, short cylinders of body which are fed onto plaster-of-Paris moulds, carrying the profile of the top of the disc insulator which this process produces. The convoluted underside is formed by a rotating metal tool, previously surface-heated by a gas jet (Fig. 3.4).

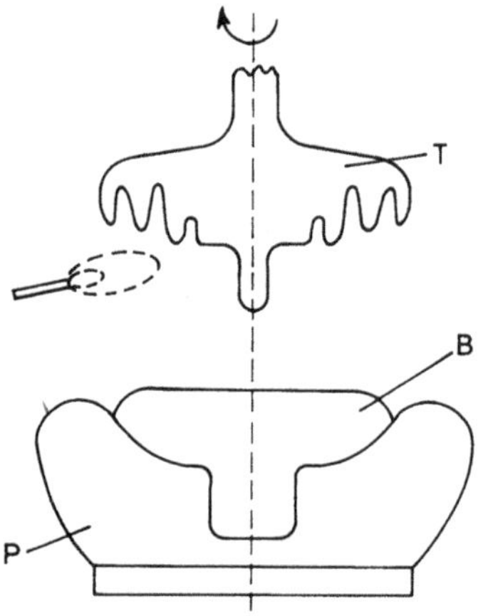

Fig. 3.4 *Principle of hot pressing*
P: Plaster mould
B: Moist clay-body
T: Heated, rotating press-tool

Higher water contents are needed for hot pressing than for turning, as stated above, not simply because of the larger volume of body requiring to be deformed but also because of the need for the shaped piece to detach itself, by shrinkage on drying, from the plaster mould. The mould and contents are passed through a warm-air tunnel, typically for two or three hours, which enables the part-dried piece to be removed for shaving and sponging, so as to adopt its required dimensions with suitable surface finish. The piece is then dried, down to 1–2% water content, in a second tunnel, a process requiring some 24 to 48 hours, depending on the size of piece.

For turned parts, either posts or hollow cylinders, the reduction of the 20% water content, which the piece has on the lathe, to the 1–2% needed for firing is likely to be both slower and more difficult than for pressed discs. Some types of defect, such as S-cracks on the axis of post or rod insulators, may originate from improper final drying: these cause failure in the kiln or on mechanical test. Electrically aided final drying is now being introduced in which direct resistive heating by passage of alternating current has the desirable effect of producing a thermal gradient from axis to wall, thus promoting orderly migration of water, for removal as vapour from the surface. Dehumidification of the ambient air, instead of or as well as heat, is also sometimes used: electro-osmosis, using DC, is being developed.

3.4 Glazing and sanding

Effectively all outdoor insulators of porcelain are glazed, although only some are sanded. There is no question, as is sometimes believed, of the glaze's acting as a water seal for the protection of a porous interior, as is the case for earthenware vessels of certain types; all electrical porcelains are fully vitrified and impermeable. Glazes confer smooth surface finish, for reduced catch of dirt and minimum specific surface area, specified colour which is commonly brown or grey, and significant increase in mechanical strength when they are arranged to be under compression. Some glazes are made semiconductive, either for control of surface electric stress and abatement of radio interference or to improve electrical performance under pollution.

Sanding is the attachment of multiple projecting grits or grains to the surface of an insulator for the purpose of increasing the adhesion of cemented-on metalwork. It finds application in some designs of straight-headed disc as well as in many substation insulators like posts and housings. Great skill and care both in the formulation of the grit itself and in its attachment, by glaze, to the substrate porcelain are needed if mechanical-stress concentrations are to be held at acceptable levels.

Glazes are glasses of complex composition. Two are sometimes separately applied to an insulator, one of specific expansibility for conferring mechanical strength to vital parts, like the head of a disc or the ends of a post, the other of different properties altogether which contains the colourants or stains. The staining elements, like Fe, Cr or Mn, although present in proportions as small as 9%, promote the formation of crystalline forms, like spinels[54,55]. Some of these may raise the expansibility sufficiently to spoil the compression effect in the glaze.

The glaze raw materials, felspar, china stone, quartz, clays, Bentonite, alkaline earths like lime, oxides of Ba or Zn to act as fluxes and the stains, are ground together in water to give about 70% particles below 10 μm. The aqueous suspension may be applied to the piece by dipping, spraying or flooding from multiple orifices, commonly to a slowly rotating piece dried to 1–2% water content. Water is removed by soaking into the surface of the piece, leaving a uniform coating. Surplus dip or flood liquid is spun or drained from the edges and sometimes mopped lightly, to avoid drips. The piece is stood, to dry by evaporation, before going to the kiln.

Grits for sanding may be specially fired, for example to give fairly uniform spheroids or ellipsoids, or pulverised from larger pieces into angular shapes. Materials vary from manufacturer to manufacturer, different philosophies reigning as to the roles of expansibility and crushing strength, relative to the porcelain and cement which the grits are required to marry.

The sand band is typically applied to the end of an insulator by painting a band of adhesive glaze suspension onto the substrate, which may already carry a layer of low-expansion structural glaze in adhered powder form. The grits are offered to the adhesive, and any surplus not held by the band of adhesive glaze suspension is removed. Unwanted gritting, for example outside specified boundaries within the pin hole of a disc, has the effect of degrading the mechanical performance and must be avoided.

The glazed and sanded piece is now ready for firing, the usual preliminary to which is the stamping of the maker's code and type mark, by offset from a pad loaded with a paint which will blacken when fired in the kiln.

3.5 Firing

Complex processes, both physical and chemical, occur [56] during firing, which is performed either in tunnel kilns, with the insulators moving on carriages, or statically in intermittent-fired kilns. Pieces may be stood on refractory dusts, or suspended from a refractory structure, commonly of SiC. Gas firing is almost universal, although some electric kilns are used for special parts. The ambient atmosphere is controlled and is usually oxidising for firings up to 1200°C. Durations may be several days: the temperature/time characteristic, both heating and cooling, is monitored and controlled.

The sequence of events, as the temperature rises, is in essence loss of remaining water, decomposition and recombination of the clays and their associated impurities, formation of viscous liquids, as the fluxes (felspars) react with parts of the clay residues and silica or alumina skeleton, sintering of the solids under the influence of the glassy components.

The regimes are approximately as follows. Up to 100°C the pore water and surface-bound water is lost. Between about 400°C and 650°C organic impurities volatilise while residual carbon burns out about 650°C, carbonates and sulphides about 900°C.

At 573°C an important physical transformation occurs which produces a significant kink in the thermal expansion characteristic. This is the transition of quartz from α to β crystalline form (Fig. 3.5).

The clays lose chemically bound water from 450°C to about 650°C. This reaction is irreversible and endothermic and is accompanied by significant contraction, which may cause shrinkage-cracks in large pieces.

Although felspars alone do not melt below about 1150°C they are capable of forming eutectics with silica and with metakaolin or clay residue at temperatures of about 990°C. Reactions leading to these eutectics begin at points or faces of contact between the various particles which make up the body: clearly, therefore, small particle size promotes early and rapid reaction while large particles, such as the quartz which makes up the skeleton, will become surrounded by solution rims of reacted material which is different, especially in expansibility, from the interior of the large particle. Such internal discontinuities affect the final physical properties, especially mechanical strength.

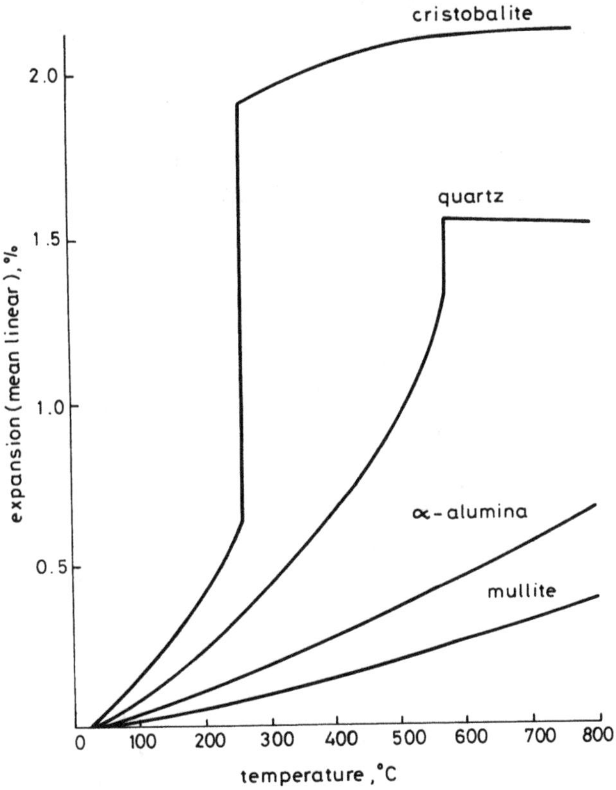

Fig. 3.5 *Expansion curves for dielectric minerals (after Reference 89)*

Above 950°C the clay residues react chemically to form mullite, an aluminium silicate, and silica, both of which take part in the fluxing reactions with the felspar. At sufficiently high temperatures, needles of mullite may crystallise out from the liquid glass and affect the final properties. Rigidity during firing depends on the viscosity of the liquid glass, governed by the proportions of Na and K in the felspar and also by the presence or absence of alkaline earths like Ca. High viscosity gives wide firing range, i.e. tolerance of temperature variations without loss of properties. It is this viscosity which allows long and heavy pieces to be hung from one end without bottom support and to survive firing without failure in tension.

As the piece is allowed to cool, the body, now a compact mass of crystals and grains bound by the glassy matrix, hardens at about 1150°C. The glaze, which has been chemically reacting itself and with the substrate body, during firing, hardens at about 1100°C. With further falls in temperature the low-expansion glaze, which is applied to mechanically important parts, is forced into compression, while the colouring glaze is required merely to follow the contraction of the body sufficiently closely not to craze or spall away. Special problems sometimes arise on cooling with sand bands, because of their physical complexity.

The principal hazards during firing are seen to arise from departure of volatiles, which must not be allowed to generate bubbles or pores, from volume changes associated with chemical reactions and tending to cause cracks, but above all from differences in thermal expansibility. The clays, quartz, alumina

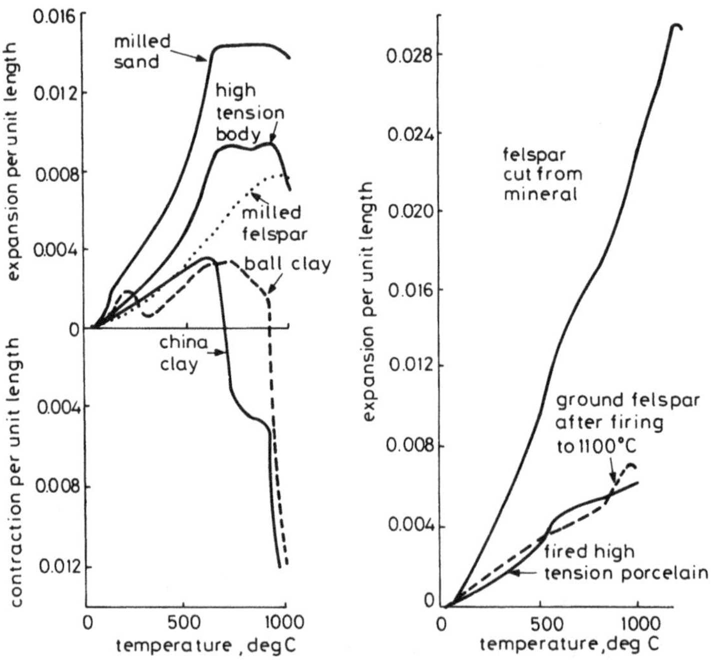

Fig. 3.6 *Expansion curves for components and porcelain body (after Reference 27)*

and felspars all have widely different expansion curves, and the composite characteristic, for the porcelain body, is dominated by different components in different regimes (Fig. 3.6). Even when the body is cooled to room temperature the internal stresses, arising from differences in expansibility, remain. In particular, the quartz grains are left in tension and constitute a potential source of failure microcracks. Removal of this mismatch, by substituting alumina for quartz, significantly increases the mechanical strength.

It is not uncommon for fired pieces to be found lacking in mechanical strength or other quality as a result of incomplete reaction in the kiln. Refiring may then

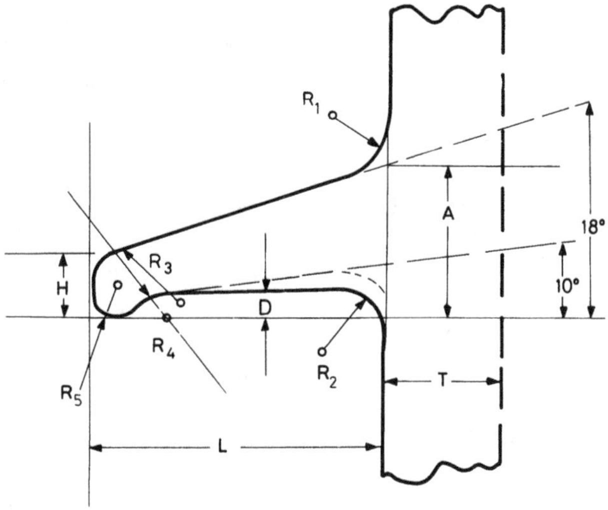

Fig. 3.7 *Limitations on shapes of barrel insulators*
For large barrels to be made by turning, some dimensions are recommended in Japan (NGK Catalogue 91) and Germany (DIN 48 115). The Japanese system relates some radii and the shed thickness to the overhang. The German system also relates the joining radii R_1 and R_2 to the overhang, but governs shed thickness by fixing the slope angles and the height H. Representative numbers are given for the two systems.

	Japanese		German	
L	50	80	50	80
R_1	12	15	15	15
R_2	10	10	10	14
R_3	15	21	15	15
R_4	5	7	4	4
R_5	5	7	5	5
A	26	40	–	–
H	–	–	13	13
D	5	5	4	4

(all in mm)

be practicable as a corrective step, especially for large pieces too potentially valuable to be scrapped.

The changes in physical strength of the body during firing and the thermally induced stresses which arise during cooling have decisive effects on the choice of shapes which can be made, with acceptable losses during manufacture. Some limitations are as follows (Fig. 3.7). The trunk wall thickness is best kept above about 0·05 of the inner diameter, to avoid deformation or bending: values between 25 and 60 mm are common. Shed thicknesses are best kept comparable with those of the trunk, to avoid contraction cracks, and radii at the joint are best kept above 10 mm for similar reasons. Sharp edges and variations in wall thickness, especially abrupt ones, are also potential stress raisers.

Since all projecting sheds present cantilever loads at their junctions with the trunk, some risk of sagging during firing always exists. For this reason, the overhang must be restricted – values of 120 to 140 mm are rarely exceeded – and a substantial section of shed must be provided at the root. A 20 mm trunk thickness is needed for an overhang of 25 mm, rising to more than 35 mm for an overhang of 90 mm. Root thicknesses between 20 and 40 mm are generally needed for walls between 20 and 60 mm thick.

Related limitations arise on the degree of complexity which can be introduced into the design of the ribbed undersides of sheds. Ribs add to the bending load at the root: they are difficult to make by turning when the body is 'green', and are likely sources of thermal stresses on cooling. These difficulties were solved, to a great extent, by the multiple-cone system, whereby modules are fired separately and cemented together afterwards.

The support in the kiln of very large cap-and-pin bodies is also difficult. The ribs-down attitude is liable to damage the contact area; the ribs-up attitude may cause mehanical weakening in the head, which is the highest stressed part in service.

It is this class of process-related limitation, on minimum achievable weights and shed thicknesses and on shapes, which has added attraction to polymeric designs.

3.6 Finishing processes

The fired piece is inspected and checked for dimensional accuracy. Discs are sometimes hydraulically pressure-tested, bursting pressure being applied inside the cavity to detect any pieces having flawed or otherwise weakened heads. Large pieces, which have been suspended from core extensions or which have their bases marked from standing in the kiln, are cut to size with diamond saws. Where necessary, ends are ground to tolerance, although grinding of hard materials like porcelain is expensive and relatively time consuming.

Metal fittings (Tables 7.1, 7.2) are generally fixed with Portland cement, either neat or as mortar with mineral additions. For disc insulators the internal pin or bolt is coated with a bitumen or related layer the functions of which are corrosion protection and high-pressure lubrication, to allow the conical part to slip under load. Caps for disc insulators and flanges for posts or cylindrical insulators are sometimes similarly coated, and even the outside of the porcelain head itself.

Precise location and elimination of bubbles by vibration are essential features of the cementing process. Hot curing of the cement, typically in steam for two hours followed by immersion in water at some 30°C for 24 hours, accelerates attainment of full mechanical strength. Insulators for tensile duty, such as discs or longrods, are hydraulically pulled to proof-load before final test and inspection.

Electrical testing may be done at this stage – finished discs are typically energised to a voltage giving intermittent external flashover, say 80 kV for a 10 kV disc, while hollow insulators will have similarly large electric intensities applied by means of chains or water electrodes, but in this case radially – or in some factories before the porcelain is assembled. The test is merely a means of detecting flaws; the applied electric intensities are well below the puncture strengths of ceramics, or indeed of any other solid dielectric. Even so, electrical testing of hollow porcelains by means of electrodes which make intermittent or point contact, especially when the test is prolonged, has to be done cautiously. Air, the immersion medium, breaks down before the porcelain: surface discharges result, with excessive field intensities at their tips. Punctures may thus occur in pieces which are not significantly substandard.

Specific tests, for behaviour under impulses, or under rain, in fog or during thermal cycling, are sometimes called for, as are impact tests and measurements of radio-interference generation at voltage. These are dealt with in Chapter 12.

3.7 Other porcelain processes

The whole subject of ceramics is at present under rapid evolution, with important developments arising from advances in materials science and from the availability of new substances. Isostatic pressing is already in use for insulator manufacture, eliminating many steps in the process and allowing close tolerances in the finished work. Machining of dry, unfired ceramic bodies is also practised, with some objections because of the health hazard from dust, as already mentioned.

Outstanding mechanical properties have already been achieved from single-oxide ceramics, some with additives such as rare earths, and fibre-reinforcement on a commercial scale must be seen as imminent.

The glass ceramics, glasses seeded and heat-treated to give controlled granular structures, offer interesting possibilities as insulator dielectrics. Their thermal expansibilities can be matched to those of metals: prototype posts, assembled from glass ceramic dishes directly stacked through metal couplers, performed well under salt pollution at the Brighton Testing Station. In principle, they could be used in assemblies which were cement-free and thus invulnerable to corrosion.

Chapter 4

Manufacture of tempered-glass insulators

4.1 Scope of manufacturing processes

The scope of the glass-insulator maufacturing processes is very much narrower than for porcelain. At present the use of toughened glass is confined to cap-and-pin insulators or those types, such as railway pedestals and multiple-cone posts, which can be assembled from disc-like modules. There is no such thing as a glass longrod, solid post or hollow shell even though, in principle, there is no reason for not applying the tempering or toughening process to such geometries.

In practice, therefore, the manufacture of toughened-glass insulators is confined to the following stages: mixing the ingredients; melting the glass; forming and heat-treating the discs; elimination of defective pieces; attachment of metal fittings. It is evident that such simplicity cries out for long runs of standard pieces, and that, when these conditions are fulfilled, cheap and good insulators may be expected.

In the following descriptions the publications of Hogg and Johnston[11, 12], describing pioneer work in England, and of Dumora, Pargamin and Parraud[58], covering the recent developments in France which have dominated the technology worldwide, have been heavily drawn upon. Reference to the original sources is recommended (Fig. 4.1).

4.2 Preparation of glass

The glass is melted continuously in a large tank furnace holding as much as 1300 tonnes. The raw materials, typically silica (57),limestone (9), Dolomite (11), felspar (4), soda ash (14) and salt cake or sodium sulphate (6), where approximate percentages are in brackets, are intimately mixed and introduced to the 'melting end' of the furnace, on top of the existing melt. The temperature may be as high as 1500°C, to contain which a highly refractory furnace lining of zirconia or similar oxide is needed.

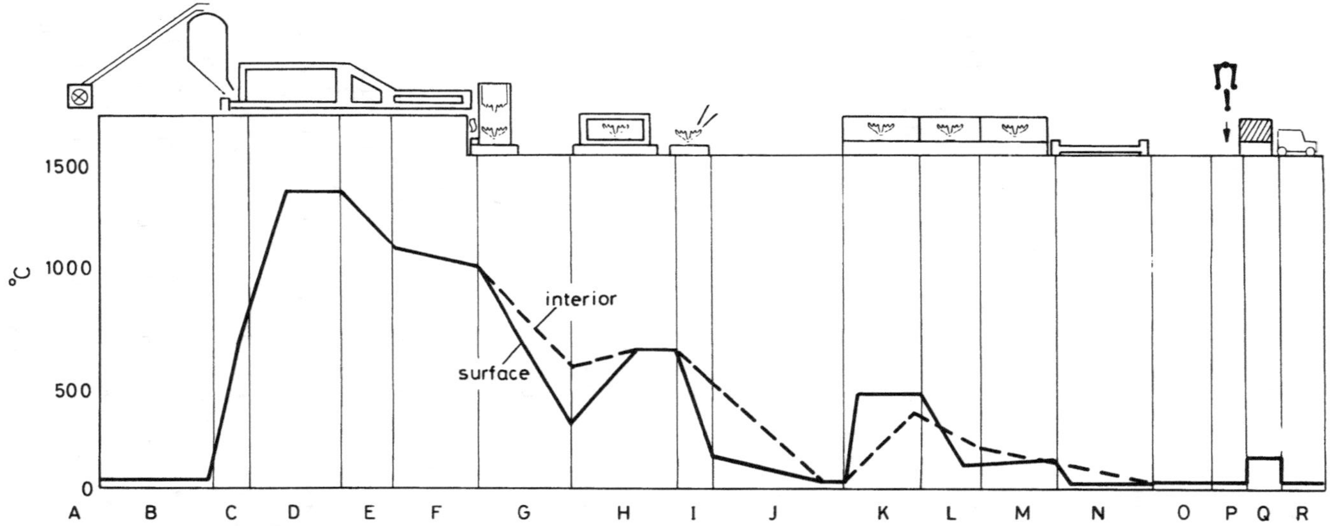

Fig. 4.1 *Manufacture of toughened-glass insulators*

A Preparation and mixing of raw materials
B Holding silo
C Charging furnace
D Melting and homogenisation
E Cooling
F Conditioning: delivery of gobs
G Moulding in chilled forms
H Equalisation of temperature
I Air-blast toughening
J Cooling of bodies
K Cold to hot thermal shock
L Cooling
M Equalisation of temperature
N Hot to cold thermal shock
O Storage
P Assembly of metal fittings
Q Steam cure of aluminous cement
R Packing and despatch

The chemistry of the raw materials under increasing temperature is related to that of porcelain. Loss of surface water is followed by decomposition, yielding oxides of sulphur and carbon as well as chemically bound water, and then by liquefaction as eutectics form between the fresh constituents and those already in the melt. Again as with porcelain, crystallisation may occur from the melt, as of calcium silicate. Oxides of alkalis volatilise and some of the furnace gases dissolve; the resultant bubbles would be undesirable in finished pieces and their removal is expedited by a fining process. This comprises elevation of temperature, to reduce the viscosity of the melt, and sometimes also the addition of specific materials[12].

The mass of the melt is in convective flow, which is desirable in permitting the fining process to eliminate both bubbles and other local inhomogeneities, but hazardous in stripping and circulating solid material from the walls of the tank. Such small fragments of refractory oxide as are entrained into the glass, and pass into the shaping operations, may cause mechanical and electrical weakness in the finished piece.

Before leaving the furnace the glass is brought to the correct temperature to form, at the end of the exit feeder, a calibrated drop or gob of glass. This falls into the mould, which has been coated with release agent.

4.3 Moulding and toughening

The molten gob is forced to flow between the upper and lower parts of the metal mould, which is often multi-piece, permitting complex three-dimensional shapes to be both formed and extracted from the mould. After ejection from the mould the surface of the piece has been cooled by conduction much below the temperature of the interior. A reheat or homogenisation is performed before the toughening, which is by carefully controlled air jets.

As stated in Chapter 2, the temperature after reheat is above that at which the glass acts elastically: the behaviour is that of a highly viscous fluid. The surface is converted to an elastic solid by the chilling air blast. The final distribution of stresses in the glass, once the whole piece has fallen to room temperature, is from a surface compressive stress to an internal tensile stress of about half the surface value (Fig. 2.9). Mean strengths are of the order of 200 MPa, (flexural).

Rejection of wrongly toughened or otherwise failure-prone pieces is done by thermal shock. The disc, at or near room temperature, is placed in a kiln at some 550°C, where it remains until the transient temperature gradient within it has reached a maximum, thus enhancing the internal tension and causing defective pieces to shatter. Some discs are partially stress-relieved at 450°C, to diminish slightly the internal tension following this cold-to-hot shock, but the return of the piece to ambient temperature is always done rapidly, to give a second hot-to-cold thermal shock, again to weed out defective pieces.

Discs are visually inspected for flaws and subjected to dimensional checks before going for assembly with metal fittings. Aluminous rather than Portland cements are favoured for glass insulators; curing of the cement is usually completed in some hours under water. Because the thermal expansibility of insulator glass is fairly close to that of the metallic fittings, there are fewer inherent difficulties with glass than with porcelain, in obtaining good resistance to thermal cycling tests. Loss of mechanical strength at low temperatures has also been claimed to be less for glass than porcelain, for related reasons.

Chapter 5

Fibrous cores for polymeric insulators

5.1 General principles

Broadly speaking, the story of fibrous cores has been one of disappointed expectations [35]. The extraodinarily high ratios of strength in tension to size and weight, which fibrous composites offered and which were seen as spelling the death sentence for conventional strings and rods, have progressively been discounted, as unfortunate disadvantages have appeared. Fibrous materials have been found highly vulnerable to deterioration under electrical and mechanical stress in outdoor conditions of humidity and pollution. Adequate levels of protection against invasion, especially near terminals, have been found most difficult to achieve, and the progressive losses of mechanical strength, mentioned in Chapter 2, have been embarrassing in practice.

It now seems likely that polymeric insulators using fibrous cores will be confined to applications where their special merits outweigh their intrinsic limitations. Such an application, foreseen by F. H. Proctor and the author more than 10 years ago [49], is for insulation at megavolt levels. Large bundles of conductors are then electrically essential for stress limitation, and mechanical loads rise above what can easily be accommodated by acceptable strings in glass or porcelain. The fibrous core, developed for this UHV insulator, was based on laminates of polymer and glass fibre. A rectangular-section beam was used in which the ends could be reinforced with cross fibres, to accept the shear loading within the mechanical termination: (Fig. 5.1). Loads exceeding 100 tonnes could easily be handled by this construction which also proved largely invulnerable to handling and electrical damage. Several of the insulators survived exposure of their cores to extreme saline pollution, under electric stress and cyclic loading, without apparent damage; no core of that construction ever failed mechanically.

It has been in the use of parallel-fibre rods that most difficulties have arisen. The underlying causes have been in the nature of the glass fibre itself, in the sizing and application of linking agents to the fibre, in the choice of matrix polymer, in defects like fibre kinks arising during the pultrusion, in the arrangements made to couple the ends of the core mechanically, in the sealing of these ends and in the handling of the completed core assembly and finished insulator.

These matters have not appeared in an orderly way, but have been identified painfully from failures in different circumstances. In particular, those faults caused by propagation of damage from an end defect have been found extremely sensitive to voltage and have generally appeared first at the higher transmission ratings, of 275 kV and above; related failures at lower voltages have taken so long to appear, in some cases, that false confidence has been generated.

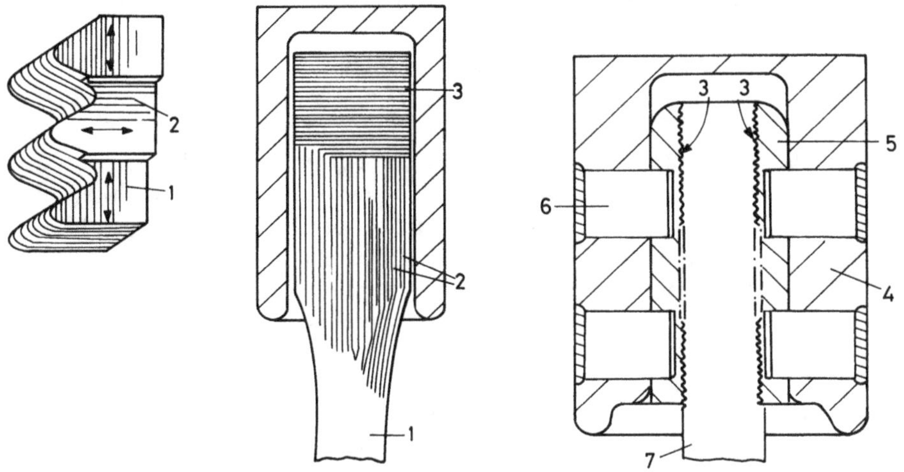

Fig. 5.1 *Termination for cross-fibre RBGF core*
1 Unidirectional fibres for carrying tensile load
2 Cross-reinforcing laminae
3 Corrugations, normal to planes of (1) and (2) and matching those on plates (5)
4 Metal housing
5 Plates with corrugated faces
6 Pins into blind holes in (5)
7 Core

5.2 Glasses and surface treatments

The most generally used glass in fibrous cores, E-glass, was apparently developed with electrical uses in mind by reducing the alkali content and substituting boron in an otherwise normal lime-alumina-silicate formulation. This glass was found by Bradwell and Wheeler[33] to be subject to attack when embodied in railway-insulator types, but satisfactory lifetimes were obtained by means of a change in the surface treatment.

Brittle fracture failure, in the form of a planar cut normal to the core axis, was apparently first observed by Proctor and Looms in a 275 kV insulator exposed to contamination in South Africa. Later work[59] found that similar failures could be produced by subjecting fibrous rods to combined mechanical stress and strong acid. Large variations in time to mechanical failure were observed between different rods using various epoxy and polyester resins as matrices[60]. Reynders had found that organic acids, which could plausibly be explained as arising from discharge activity in the polymer, were also capable of causing brittle fracture failure. His experiments also showed that other formulations than E-glass could give longer lives in contact with acid[61]. More recent work by de Tourreil and others[62] has indicated that reduction of the boron content in E-glass and substitution of other radicles, like Mg, Zn or Ti, gives improved resistance to acids, and that oxalic and nitric acids behave in different ways as aggressors.

Much of this work lacks practical relevance. The concentrations of acid which, in conjunction with realistic values of mechanical stress, are needed to produce failure in reasonable times are very high – of order 1-normal, far higher than would be expected from natural sources or from corona discharges – and cannot be deemed probable in normal insulator service. It seems likely that electrochemical effects, such as the production of H^+ ions by electrolysis of leaky surface pollution, have to be invoked. In such a case it cannot be regarded as proven that the adoption of a new glass, or treatment, or resin, which shows improvement in strong acid will also show comparable benefit in real service.

When it comes to control of process variables, such as good wetting of the glass by polymer, avoidance of cavities, preservation of parallel and kink-free fibres, full cure of resin and elimination of monomers and residues, little can be done apart from relying on conscientious manufacture. Internal faults, which will only later develop into electrical or mechanical weaknesses, are almost impossible to detect by normal quality-control measures.

The questions of mechanical coupling and of damage by flexure and torsion are dealt with in Chapters 2, 7 and 8, but the most important safeguard, adequate sealing against both water and contaminants, deserves full treatment here.

5.3 Sealing of core ends

Although in principle cores may be humidified by the inward migration of water vapour radially through the walls of the housing, this is certainly not a common cause of failure in practice. As mentioned in Chapter 2, silicone elastomers, known to be highly permeable to water vapour, or the thin layers of highly filled epoxy resin, also permeable and widely used in the earliest polymeric insulators, never seem to have caused core failures without some form of contributory defects like cracks or splits, which would admit *liquid* water. Sealing of the ends of the core, where conditions are severe and unique, against water in any form appears vital.

The metallic end cap itself should preferably be free of holes or seams: one design, made from compressed tube, allowed sufficient migration through the unwelded faces to cause failure, even though the gap was almost undetectable.

The sealing arrangements must also take account of the fact that some degree of pumping is common within end fittings, both from cyclic mechanical loading of the core and from thermal excursions.

At the mouth of the fitting, where core and housing enter metal, seals of varying complexity have been used, including double O-rings on railway types[63]. Polymeric sealants have included caulks, based on silicones of room-temperature vulcanising formulation, mastics[64], hot-melt adhesives and metal-loaded paints[65]. An external hood (Fig. 5.2) might seem a useful water barrier, but is in fact undesirable. The shoulder of the metal fitting produces a stress concentration while the leakage current develops a voltage drop; these effects combine to produce puncture of the hood and an entry port for liquids, which, moreover, is at the very place most likely to be contaminated with electrochemical products of electrolysis produced by the leakage current.

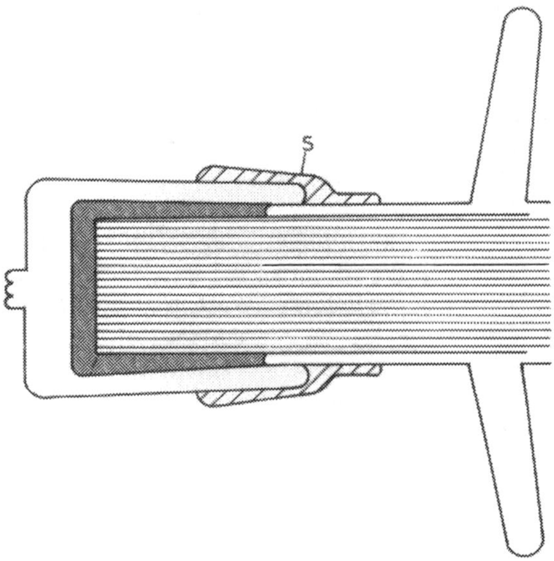

Fig. 5.2 *External hood at metal/polymer transition*
Puncture occurs at metal shoulders S, allowing invasion by pollutants at vulnerable region

Especially on vertical insulators, the design of the fitting and seals must prevent the accumulation of a pool of water. Where this precaution was neglected[49] heavy attack resulted, again by electrochemical products, both on the metalwork and the polymeric housing: the deteriorated seal acted as a retaining sponge for the chemical aggressors.

5.4 Service experience with fibrous cores

Data on failure rates of products are never easy to collect and rarely capable of expression in significant ways, but some facts on service experience with fibrous-cored insulators have been published by manufacturers and users.

Armstrong[66] gives data on one type of core which was protected by mobile silicone treatment of the interface with the housing. Some 175 000 insulators, both of suspension and post type, were installed between 1976 and 1983, of which seven cores were admitted to have failed. Proctor[67], reporting on 12 000 epoxy insulators installed between 1972 and 1984, quotes eight faults from an early batch of 3000, but none from a later design with improved water seals. A working group of CIGRE, which assembled data from several countries on core failures, quoted small numbers from cores housed respectively in fluorocarbons and silicones, and none from cores which had been vulcanised to EPDM sleeves.

Only approximate values of annual failure rate can be deduced, but these seem not far out of line with the numbers relating to glass discs or porcelain longrods, which are quoted as lying between 0·1 and 1·0 per thousand annually[68, 69].

More disquieting information comes from experimental insulators for megavolt-system experimental lines and from recent failures of temporary supports used in live work at 275 kV. One UHV insulator failed after only hours under voltage, while the foam-filled temporary support had had a total lifetime under voltage of only some 100 hours, although its total age was several years and the conditions in which it had been used were likely to cause mechanical damage. The consequence of the first failure was to drop a 1200 kV bundle into the head of its associated tower; the consequence of the second was a close approach to killing a live-line team of workmen. Unforeseen hazards of this nature are bound to influence prudent designers in favour of proven, if inelegant, conventional insulators.

Chapter 6

Polymeric housings

6.1 Relationship between shape and material

A conventional assumption that any shape can readily be formed in polymer is incorrect and misleading, as far as housings for polymeric insulators are concerned. The two common requirements in a practical transmission insulator, lengths exceeding one metre and profiles containing re-entrant sections, are virtually impossible to meet at acceptable cost in any polymeric material which would offer a reasonable life expectancy.

It is instructive to compare a porcelain longrod with possible alternatives in polymer: this is, in fact, the only class of insulator in which comparison can be made, since most others, such as bushing shells, substation posts, cylindrical housings or switchgear interruptor heads, are practically never made in polymer.

Lengths up to 1·5 m are routinely made in porcelain, with a choice of profile limited only by what can be turned in a lathe, including re-entrant and complex forms. Variations in length are readily accepted by the manufacturing process without excessive extra cost. Now turning to polymers, a similar length in one piece can, at present, be made only in a cast thermoset (although new processes are nearing realisation for other polymers, at the time of writing). The shape is limited to what can be extracted from the mould, which, in practice, either confines the choice to profiles which are simple and not re-entrant or calls for moulds which can be dismantled and are therefore slow to use in production. Mould costs are very high: there is, in practice, a large economic penalty in changing the length of the product, even slightly, to meet changes in requirements by the user.

Further, if it is required to make the housing in a material which cannot be cast from liquid, for example in an elastomer which needs high press capacity, monolithic long bodies become very expensive: an assembly of moulded parts, with perhaps a preliminary extrusion to cover the core, is then dictated. Such an assembly will comprise some tens of radial joints between the separate moulded parts, each joint representing a possibly weak point for invasion by pollutants.

As mentioned in Chapter 2, polymers which are chosen for use in housings are often those which are intrinsically non-adherent: their freedom from unwanted contamination is bought at the price of inability to be easily bonded to neighbour or to substrate.

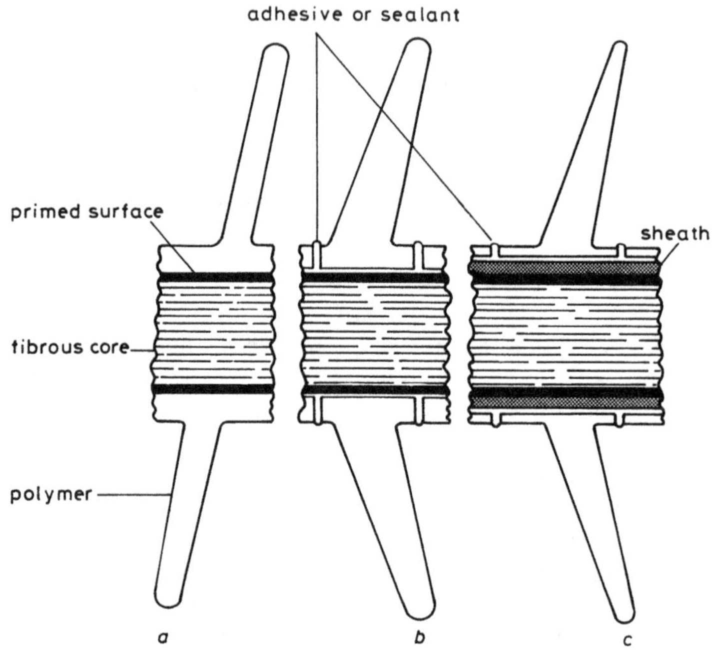

Fig. 6.1 *Make-up of polymeric housing*
a Housing moulded directly onto core
b Multi-part housing. Adhesive or other sealant between parts, parts and core
c Core continuously sheathed. Parts stuck, bonded or vulcanised to each other and to sheath
*c** Fluoropolymer housing is as (*c*) but sheath is replaced by sealant

For a polymeric housing, therefore, the general rules are as follows (Fig. 6.1). For one-piece housings the design of profile must be simple and cannot embody re-entrant parts. For multi-piece housings the creepage path must comprise at least two different polymers, respectively the moulded-part and the bonding-layer constituents. It is unreasonable to expect identical electrical and mechanical performances for these constituents. Invasion at joints is likely to be serious, and statistically must be expected, since each unit comprises many joints.

In one class of polymeric housings bonding is not attempted, either to the core or to neighbouring moulded parts. Instead, layers of silicone-based mobile grease or oil are used, both on the fibrous core and between the moulded parts. Clearly, such a design (Fig. 6.2) is heavily reliant on the migration rate of the mobile grease and on the feasibility of preventing loss of grease and avoiding invasion, as by high-pressure washing water. In general, however, a combination of extrusion and bonding is relied on.

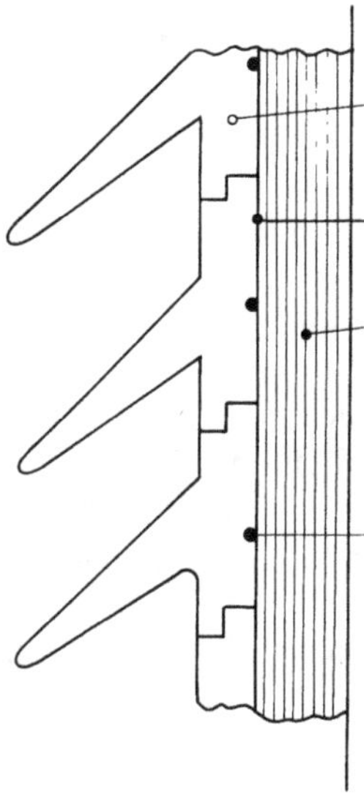

Fig. 6.2 *Polymeric insulator based on greased RBGF rod*

6.2 Extrusion and bonding

Casting from liquid thermoset resins has been mentioned, to which was related the earliest manufacture by sequential casting from room-temperature-vulcanising (RTV) elastomers.

This system used massive pieces of silicone elastomer, each part being allowed to bond to its neighbour during cure. Another system used sintered sections, each moulded from powdered fluoropolymer, such as PTFE or equivalent, and bonded to its neighbour by a fusible fluoropolymer of different formulation. The intrinsically non-stick assembly was internally sealed to its fibrous core by means of an adhesive of different composition, usually an epoxy (Fig. 6.1*c**).

Following pioneer work in France[70], however, the most favoured system has been a combination of primary protection for the fibrous core, by means of a heavy cylindrical sheath of elastomer, applied by extrusion and vulcanisation, and a set of separately moulded shed sections, usually but not always of similar composition to that of the sheath. The long established and highly developed technology of the vehicle-tyre manufacturers has been applied here to considerable effect, and units suitable for duty at 765 kV have been successfully made in single unjointed assemblies. Some disadvantages have appeared which are discussed in Section 6.5.

6.3 Casting and moulding

One-piece casting, in alicyclic epoxy resin loaded with fillers, has been successfully achieved on fibrous cores, both laminated and pultruded, to give monolithic insulators suitable for use up to 400 kV system[71,72]. Especially with pultruded cores of small section, such as the 380 mm^2 which was used in early designs for 275 kV duty, it proved difficult to hold the core free of sag curvature under casting, and to remove the cured assembly from the very complex mould without damage. The parallel inclined planar profile which was adopted to allow removal from the mould proved, both under test and in service, to give a good electrical performance (see Section 6.6).

Cast resin housings cannot readily be made elastic: some epoxy housings have been based on so-called flexibilised resins, of a dough-like consistency which allows some accommodation of core extensions, bending strains and the like, but such formulations have serious deficiencies in comparison with elastic rubbers, and have proved to be subject to failure by cracking. The technology of casting large volumes is also a very difficult subject since the curing reactions are mostly exothermic; consequently the avoidance of thermally generated internal stresses either calls for very long curing times or for some type of compensation like inward heat feed.

Moulded housings have been made integrally with cores for small distribution-voltage insulators, where core and sometimes end fittings can be placed in the press to receive the injected elastomer under high pressure. For the reasons stated above, the costs of using similar processes for transmission insulators have, so far, proved to be prohibitive, and virtually all the large moulded housings are assemblies. The modules may be butted or arranged to interlock: some designs incorporate internal storage reservoirs for protective greases (Fig. 6.2).

Because of the good recovery of most elastomers from deformation, some degree of re-entry is possible in moulded profiles. However, the high degree of filler loading which is needed to confer discharge resistance, especially on polyolefine rubbers, does restrict the available complexity of shape. Mould-release agents are almost always needed to allow the module to be extracted after cure. These are hydrophobes which are unobjectionable on outside parts of the housing, but which seriously complicate bonding to neighbours or substrates, unless scrupulously removed beforehand.

6.4 Other fabrication processes

The range of processes whereby polymers can be formed into complex shapes is increasing almost daily, and some apparently unlikely candidates have been successfully adapted, at least on a trial basis, to the making of insulator housings.

The experiments of D. H. Lucas[73], who assembled multiple cones, made from thin sheet by blowing or vacuum forming, as skirts onto a stem, were particularly interesting. He provided enormously long creepage paths by close packing of graduated cones (Fig. 6.3), using his own 'precipitator window' principle[74] (that dust cannot migrate to the end of a cavity which has a large ratio of length to cross-section) to keep the stem unpolluted. His steep-angled cones, based on the 'easy grease' shape developed by the author's group[75], gave good water draining.

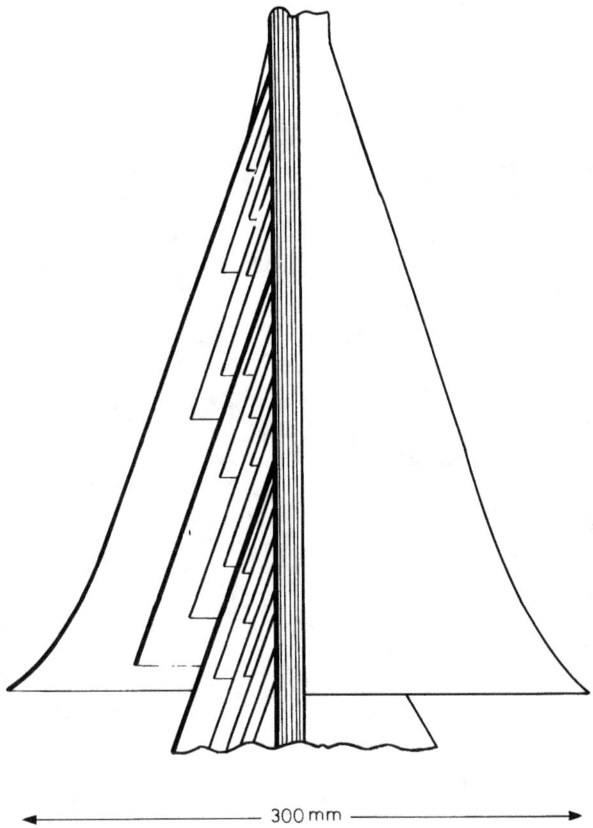

Fig. 6.3 *The Lucas insulator*
Long creepage, about 10 × axial length, allows use of poor polymer. One of four cascaded sections is shown, for 132 kV test piece

The outcome of these inventive steps was an insulator which, on test, carried no detectable leakage current and showed no damage by tracking or erosion, in spite of subjection to intense marine and ash contamination and weathering. The latter facts were especially significant since his choice of sheet material was polystyrene, perhaps the worst conceivable polymer and most liable to deterioration. The prototype later failed elsewhere because of clogging by bulky cement and dust contaminants, but the principle – that intrinsically unsuitable materials and assembly methods can be offset by long creepage – remains established.

The development of elastomers which retract to predetermined shapes under heating has opened another door to the making of polymeric housings. Heat-shrinkable tube, separate shed assemblies and continuously finned sleeves have been applied to fibrous as well as ceramic substrates. These have performed well, when proper arrangements have been made to protect the interfaces, and have raised the possibility of very long, one-piece insulators of minimum weight and maximum creepage length[76] (Fig. 6.4).

Fig. 6.4 *Optical-fibre conduit*
Multiple optical fibres are carried along an RBGF strength member, immersed in viscoelastic sealant under retractable polymeric housing. 400 kV rating. (photograph ack. Raychem UK Ltd.)

6.5 Behaviour of polymeric housings: tests, trials, service

Two aspects are considered, respectively the ability of the housing to resist flashover and its effectiveness as a protection of the internal parts and, in particular, the fibrous core. The rates of deterioration of both functions depend on the electrical and mechanical stresses and on the severity of pollution and its nature.

On flashover performance, a housing may either suffer changes in its surface which lead to increased electrical conductivity in a given pollution or become undermined or channelled in such a way as to diminish the effective creepage-path length.

Very large changes occur with weathering and under attack by discharges in some housing surfaces, both short term and long term. Especially for cast resins which contain mineral filler, discharges and weathering cause severe roughening and exposure of filler grains[77]: the consequence is a large increase in leakage current and corresponding decrease in flashover voltage. This has led (Section 2.5) to near-halving of performance after periods in service of only one or two years. Superimposed on this slow but easily explicable decline are quite rapid temporary changes (Fig. 6.5), the cause of which is not known but which may well be related to enhancement of surface free energy by action of corona on the polymeric component of the surface. Corona activation is a standard processing aid when inks or paints have to be applied to hydrophobic polymers like polyolefines.

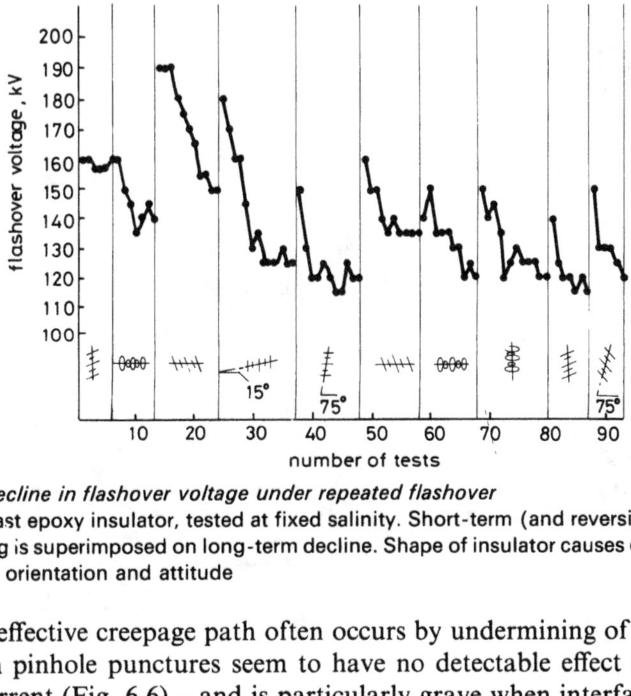

Fig. 6.5 *Decline in flashover voltage under repeated flashover*
Cast epoxy insulator, tested at fixed salinity. Short-term (and reversible) conditioning is superimposed on long-term decline. Shape of insulator causes complex effects of orientation and attitude

Loss of effective creepage path often occurs by undermining of sheds or fins – although pinhole punctures seem to have no detectable effect in increasing leakage current (Fig. 6.6) – and is particularly grave when interfacial invasion extends so far that puncture of the housing with flashover of the external residual surface can occur (Fig. 6.7). Complete rupture of the housing and mechanical collapse of the insulator are usual consequences.

External tracking is no longer an ordinary mode of failure since virtually all housings now employ either track-resistant formulations or purgative fillers

(Chapter 2). Erosion is, however, a universal phenomenon which arises from disparate causes. Direct ablation occurs at the sites of surface discharges by thermal volatilisation of the polymer. Even low-current discharges, carrying about 1 mA, cause ablation[78] because they run close to the surface and because their internal field intensities are high, and hence their powers per unit length and temperatures. Provided discharge erosion is distributed over most of the surface of the housing, its consequences are not grave and may even include some degree of self-cleaning[48].

Fig. 6.6 *Pinhole punctures in polymeric housing*
Punctures have little effect on leakage current. Close grouping shows that each is isolated resistively from others. Damage mechanism is unexplained

Localised erosion is a different matter, since this can lead to penetration of the housing or to channels in the surface which act as extensions to the electrodes. A second cause of erosion, important because it is localised and because it occurs at the highest stressed parts of the insulator, is electrochemical by-products of electrolysis of the pollution layer. Where NaCl is present, caustic soda and oxidants are produced at places where the leakage current transfers from the ionic to electronic conduction mode, i.e. at metallic electrodes and at carbon particles on the surface. Some housings are attacked in this way to such an extent that lengths of the core are exposed at the electrodes [79].

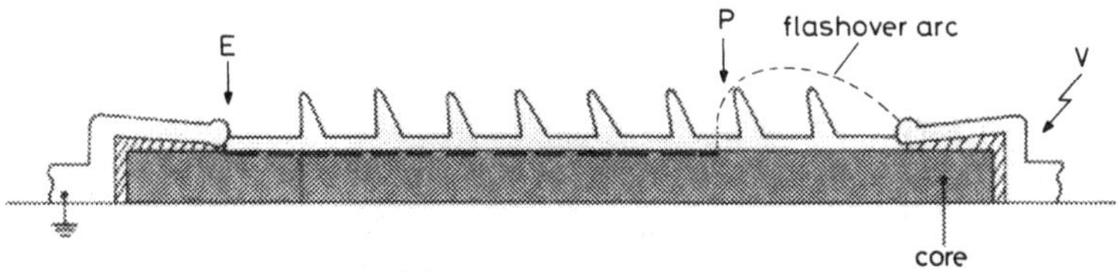

Fig. 6.7 *Mechanical failure by flashover*
Earth potential is transferred to P from E by invasion: EP may be high-resistance path. Puncture of housing at P allows flashover. Fault current (kA order) destroys insulator by bursting

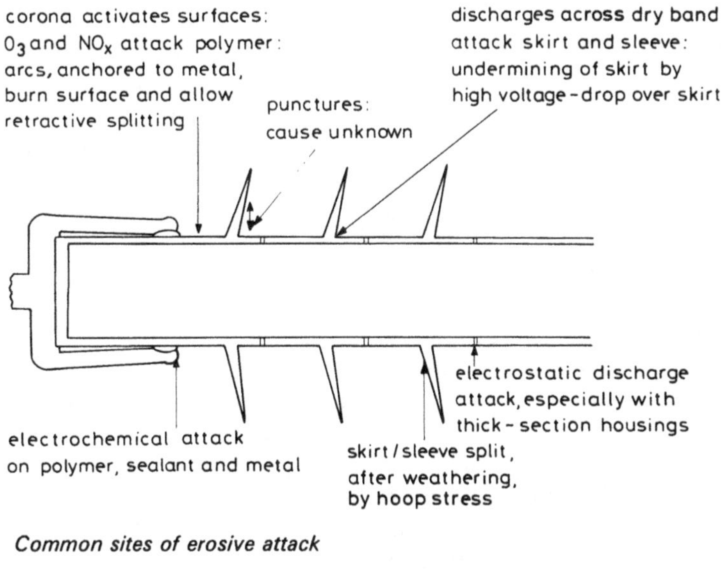

Fig. 6.8 *Common sites of erosive attack*

A third cause of erosion is burning, at foci of leakage current or associated with special pollutants, such as the metallic oxides which are present in metallurgical effluents and on railway insulators[80,81].

Common sites of erosive attack (Fig. 6.8) include those (Fig. 5.2) where housing polymer is allowed to extend over metallic electrodes. Electrostatic stress relief is often provided at the ends of transmission-type housings, by corona rings or arcing horns, to limit surface electric gradients in the neighbourhood of the terminals. Benefit is solely obtained in fair weather, since the surface resistance of wet housings overrides the capacitive effects of the rings.

A type of housing failure which affects elastomers, either stretched parts left in tension to give good grip on the core, or some kinds of heat-retractable sleeve, is tearing or cracking as a result of weathering and leakage-current attack. Details of this and other service failures are given in CIGRE reports[61,82].

6.6 Profile and performance

The profile of a polymeric housing, i.e. the creepage path length and the way in which this length is disposed, affects both the flashover voltage of the insulator and the likelihood that the housing will be damaged by effects of leakage current.

One of the few direct comparisons of profile in which only the shape was varied while the creepage length, interelectrode spacing and housing material were all held near constant, was made by Proctor and Looms and published by CIGRE[35,48]. The important conclusions of this study are summarised in Fig. 6.9 and Table 6.1. In the moderate wetting rates of the standard salt-fog test (IEC Publication No. 507, 1975), the use of radial fins, having negligible slope from the horizontal, in the profile of a vertical insulator is seen to be inadvisable. The use of sloping skirts, either planar or conical, confers an improvement in flashover voltage at a given pollution severity of at least twofold. What is surprising, however, is that a similarly large improvement is achieved for the horizontal attitude. It is concluded that 'protected creepage', as a proportion of total creepage, is significant with these uncomplicated shapes in determining flashover performance.

Evidently erosion is related to profile, since electrochemical erosion depends on quantity of leaked charge and discharge erosion depends on number and amplitude of leakage-current pulses. It is well established that the addition of even a few radial sheds or skirts to a simple cylindrical insulator dramatically reduces the leakage activity[83]. Increases in leakage path and encouragement of dry regions by the use of skirts which give protected creepage are therefore valuable features, in principle. It must, however, be realised that few very large skirts must lead to local dry bands, and hence stress concentrations, in practice: such conditions will favour localised erosion as well as undermining or puncture of the large skirts. Recent moves in profile design have therefore followed the principle of multiplicity of small skirts, preferably sloped and alternating in overhang to minimise short-circuiting by drips; thin housings can then be used because the probability of puncture is much reduced (Fig. 6.4).

On the question of thickness of skirts, sections much thinner than those used conventionally with glass or porcelain are employed with some polymers. The main reasons are two: polymers are much more costly than common ceramics

Table 6.1 *Effect of profile on performances in different attitudes*

Sample	Make-up of sheds	Flashover voltage (kV)	Salinity (kg/m^3)	Attitude
Horizontal planar sheds	23 large + 24 small	84 76	40 80	Vertical
Inclined planar sheds	22 large + 23 small	159	80	Vertical
Re-entrant conical sheds	23 large + 24 small	168	80	Vertical
HPS		98	80	Horizontal
IPS		149 average of different orientations	80	Horizontal
RCS		160	80	Horizontal

Comparison of flashover voltages between epoxy insulators of different profile but equal creepage length and electrode spacing. Creepage = 6120 mm; spacing = 1360 mm

Tested at CERL, Leatherhead, to IEC 507: salt-fog test).

Sample	Make-up of sheds	Insulating distance (mm)	Creepage distance (mm)	Test voltage (kV)	Withstand salinity (kg/m^3)	Attitude
HPS	14 large + 13 small	1,165	3,540	84	0·625	Vertical
HPS	14 large + 13 small	1,160	3,540	84	28	Horizontal
IPS	18 large + 19 small	1,130	4,870	84	225	Vertical
RCS	19 large + 19 small	1,120	5,000	84	> 225	Vertical

Comparison of withstand salinities between epoxy insulators of different profile and creepage.

Tested as CESI, Milan, as above.

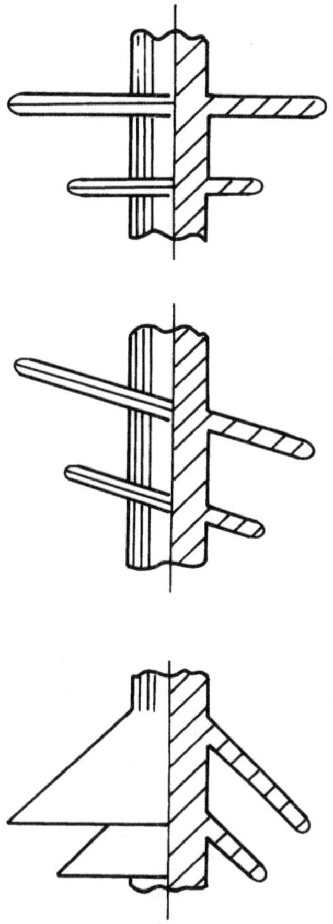

Fig. 6.9 *Effect of profile*
These profiles were directly compared in salt fog.
Material was cast alicyclic epoxy: lengths and creepages were equal

(Section 1.3), and thin parts are readily moulded or pressed in polymers. Some polymeric housings do use substantial sections thicker than 25 mm, but in these cases it must be remembered that electrostatic-stress enhancement will occur in any thin air films which separate these massive parts, with the potentiality of initiating erosion (Fig. 6.8).

Chapter 7

Terminal fittings for insulators

7.1 Terminal materials

The metal fittings which terminate ceramic insulators are almost always made from ferrous material. An exception is bushing shells, the flanges of which must not be of magnetic material. Caps and flanges are cast. The principal materials here are malleable iron, spheroidal graphitic or ductile iron. Pins, to be buried in cement within the ceramic parts of discs or pedestal posts, are more highly stressed mechanically than caps and are almost always forged from steel.

Malleable irons, either black heart or white heart, are made from cast iron by heat treatment, respectively in neutral and oxidising atmospheres. Ductile irons are made by the addition of reactive metals, like magnesium, to the cast iron, with subsequent heat treatment. The carbon separates into nodules, in black heart iron, of lamellar graphite up to 50 μm maximum dimension; in ductile iron the graphite forms spheroids of up to 35 μm diameter. Principal properties are given in Tables 7.1 and 7.2 (Figure 7.1).

All ferrous fittings are hot-dip galvanised, including the steel pins, usually made from medium carbon steels but sometimes from special corrosion-resistant alloys. Galvanisation is a sacrificial protection, the electrochemical basis of which is undermined by the passage of leakage current in pollution. For severely salt-polluted places and in DC insulators a heavy sleeve of zinc is fitted around the pin, where it enters the cement, to accept attack by leakage current without loss of mechanical strength in the pin (Fig. 7.2).

Principal properties of the alloys of aluminium and zinc which are used as materials for fittings are given in Table 7.1; copper-based alloys are also used in bushings and some traction applications.

The same classes of material are used for polymeric as for ceramic insulators. Specially ductile metals are requisite where crimping or compression is used to fix fittings to fibrous composite rods[84].

Table 7.1 *Properties of metals used with insulators*

Material	Cast iron, Black heart (MCI)	Cast iron, Ductile (SGI)	Light alloy	Steel		Steel		Diecast Zn alloy
				A	H	A	H*	
Density (g/cm^3)	7·0	7·0	2·8	11·0	11·0	11·0	11·0	7·2
Tensile strength (MPa)	300–400	350–450	300–350	570	800	650	800	260–300
Elastic limit (MPa)	200–250	230–240	220–250	330	600	360	650	250–280
Elongation at break (%)	8–18	17–18	10–14	21	12	17	10	5–8
Tensile modulus (GPa)	100–130		75	200		200		85
Expansibility ($\times 10^{-6}$/°K)	12	12	21	11		11		26
Brinell hardness	110–145	160–180	100	170	230	190	230	
Fracture energy density at 20°C (kJ/m^2)	10–12	15–17	10	50		30		

* A: annealed
H: hardened

7.2 Mechanical design of fittings

7.2.1 Fittings for porcelain and glass

Wide mechanical differences exist between ceramics and fibrous composites: these affect the design of terminal fittings. Ceramics are brittle, stronger in compression than in tension, and isotropic. Composites are non-brittle, weaker in compression than tension, and highly anisotropic. There are thus few common features in fittings, apart from the use of conical interfaces in some cases.

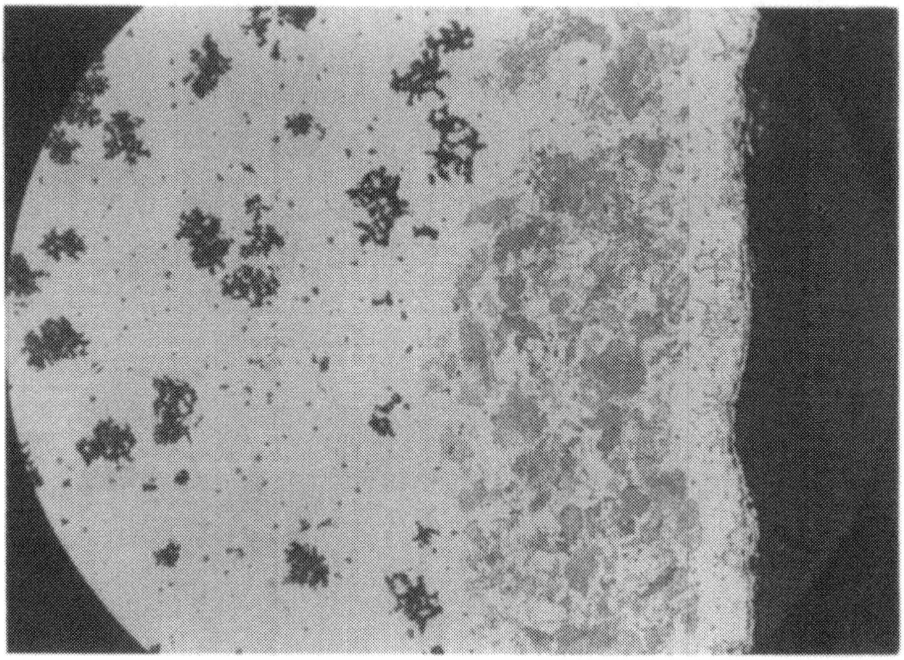

Fig. 7.1 *Malleable cast iron*
Section of blackheart insulator cap. Heat treatment leaves a rim of pearlitic iron. The carbon lamellae are about 50 μm in diameter

The design of fittings for strut insulators, such as solid posts, is relatively straightforward (Fig. 7.3). The principal need is to avoid excessive concentrations of stress in the ceramic. For cap-and-pin discs, however, and to a lesser extent for longrods, the transfer of tensile into compressive stresses between metal and ceramic is complicated.

The basic principles of disc design are presented in IEC Document 575 (Fig. 7.4) as well as in other papers, for the most part by manufacturers[85–88]. There is some disagreement on details, for example which of the interfaces require to be freed by lubrication, but the general picture is as follows. The tensile load is converted into compression by opposed conical surfaces, respectively at the top of the pin and around the rim of the cap. Load is transmitted through cement, outwards from the pin and inwards from the cap, to the glass or porcelain dielectric, which, over much of the cylindrical portion of the head, is in radial compression (Fig. 7.4).

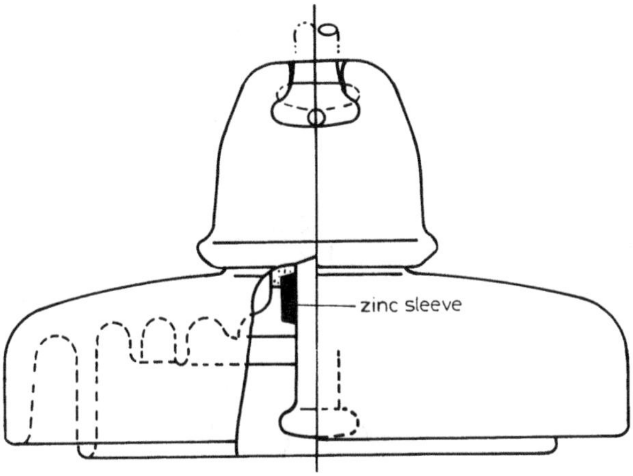

Fig. 7.2 *Sacrificial zinc sleeve for DC or severe pollution*

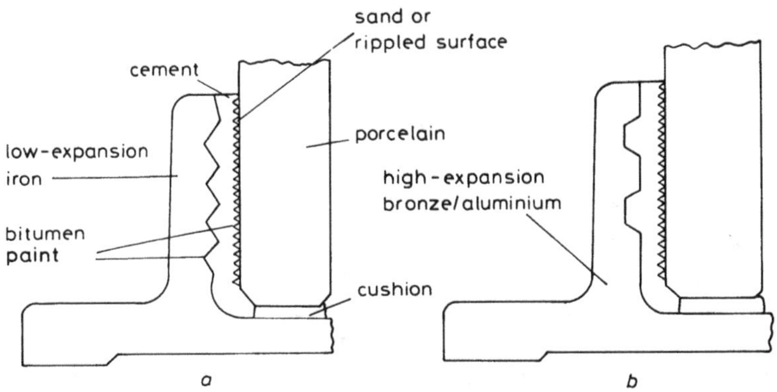

Fig. 7.3 *Fittings for strut insulators*
High-expansion non-ferrous fittings (*b*) are needed where metal carries magnetic induction.
Principal design requirement is to avoid cracking porcelain under *cantilever* loads.
a Ferrous fitting
b Non-ferrous fitting e.g. for bushings

The reasons for the complex shapes of the metal fittings and for the use of slipping or lubricated interfaces include differential thermal expansion and cyclic mechanical loading. The expansibilities, all $\times\ 10^{-6}$ per deg C, are respectively, for metal, cement, glass and porcelain, about 11·5, 10·0, 9·0, 6·0. Typical temperature excursions will be from $-30°C$ to $+70°C$, showing the necessity for allowing sliding motion by both cap and pin (Fig. 7.5). Cyclic mechanical loads are also common, especially as a result of varying wind speeds and induced mechanical oscillations, variations of 50% being known in severe cases of galloping or ice shedding of conductors.

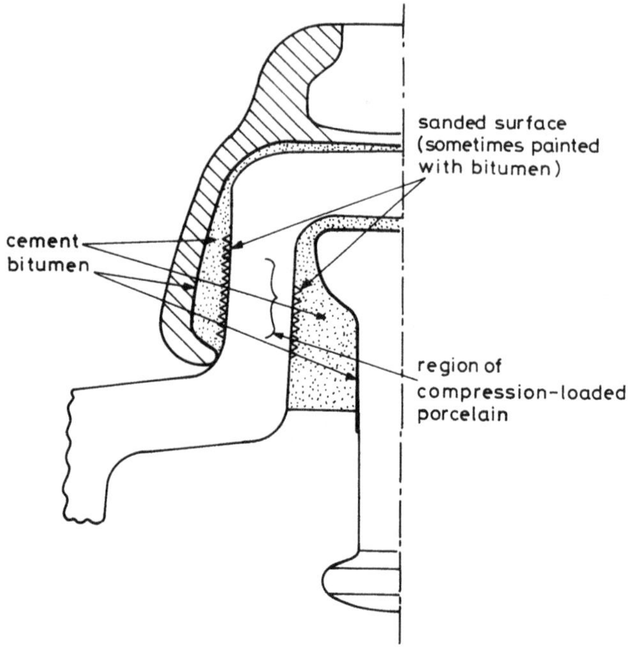

Fig. 7.4 *Basic principles of insulators using cap and pin*

The choices of cap angle and pin-head shape (Fig. 7.5) help to determine the performance under thermal and mechanical cycling. Too small a cap angle promotes wedging, i.e. irreversible movement of the cap along the cement slope: too large an angle subjects the dielectric to excessive shear stresses. Similarly, a poor shape or incorrect angle in the pin-head leads to excessive hoop stresses under expansion or heavy load. It is the practice in some designs of disc to use more than one conical face, on cap, pin or both, especially for the higher mechanical ratings (Fig. 7.6).

Computations of stress pattern are possible by means of finite element analysis, but to date poor precision has been obtained, largely because of the difficulty in establishing realistic boundary conditions[89]. These do, however, illustrate the main features of mechanical failure arising from excessive hoop stress, shear stress or principal tensile stress at concentrators like the edges of sand bands (Fig. 7.7).

Because of the greater variability in mechanical strength of the brittle dielectric in comparison with that of the metallic parts, it is common practice to design the metalwork to fail with increasing load before the dielectric. An acceptably narrow range of failure strengths, usually at least three standard deviations above the mechanical rating, is obtained in this way (Fig. 7.8).

One aspect of the dynamic performance of metallic fittings on disc insulators which has important secondary effects is that of pin extension under load. For example, on a 400 kN cap-and-pin insulator having a steel pin of diameter 28 mm, a stretch of some 30 μm occurs in the pin at one-half rated load. Contraction of the cross-section because of Poisson's ratio thus amounts to some 10 μm, which may be sufficient to allow ingress of water and thus initiation of corrosion at any unprotected part of the interface between pin shank and cement. Such corrosion will ultimately put the dielectric into hoop stress, a condition under which it is most likely to fail. Although there are greater opportunities for invasion at the edge of the cap, the consequences are relatively trivial since any corrosion here increases the inward pressure on the dielectric; i.e. it actually reduces the hoop stress.

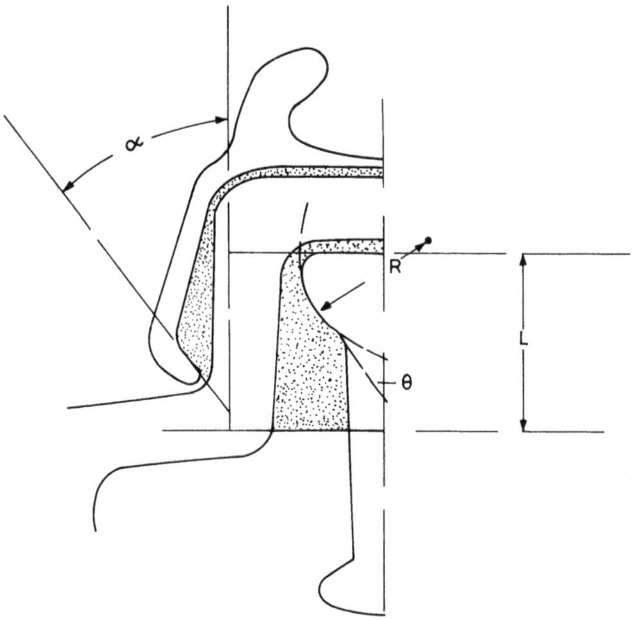

Fig. 7.5 *Shape of insulator head*
Mechanical strength and behaviour under cycling, thermal and mechanical, depend on cap angle α, pin-head radius R, pin-head angle, θ. Prospective expansions of length L, respectively in metal, cement, porcelain are, between −30 and +70°C, 53, 46, 28 μm. Pin stretches about 30 μm and section contracts by about 10 μm, under half load

For porcelain longrod insulators, the conventional terminal fittings are caps coupled to a tapered conical surface by lead–antimony alloy, sulphur sand or Portland cement. Some concentration of stress at the mouth of the cap is difficult to avoid, and fracture under destructive test usually occurs there (Fig. 7.9).

7.2.2 Fittings for fibrous composite cores

The principal application of fibrous cores is to tension insulators, for which the properties of multiple parallel fibres are most valuable. The tensile strength of a good fibrous core is such that a 22 mm-diameter rod will have the same tensile rating as a 75 mm-diameter core in porcelain: it follows that the fibrous core is stressed 12 times higher than is the porcelain and will extend nearly 20 times more, because the elastic moduli are about 11 : 7, porcelain/fibrous composite.

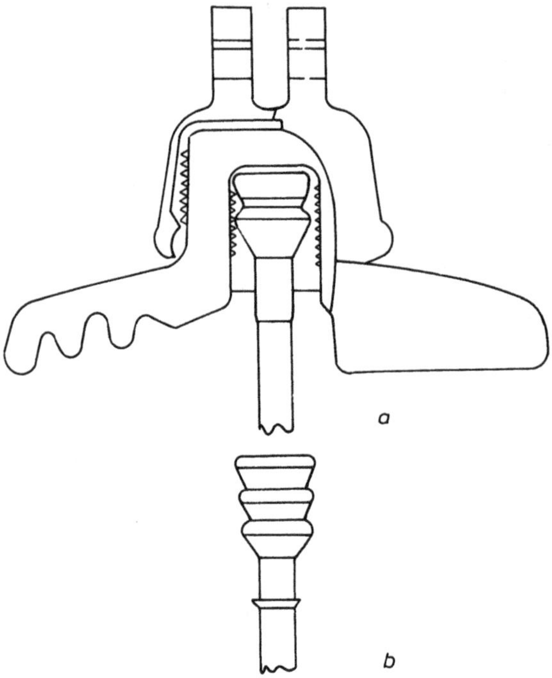

Fig. 7.6 *Multiple-angled caps and pins*
Some discs for heavy loads use more than one angled section on cap, pin or both. Two cones (*a*), or three (*b*), give much reduced tensile stress in porcelain, compared with single curved head (Fig. 7.5).
Thickened pin reduces relative strain under load.

All metallic terminations for fibrous cores must thus be designed to accept extensions under load and Poisson contractions of very large values, compared with those for ceramic insulators. They must also allow gradual progression of load from metal to fibrous composite, since the fibres are coupled mechanically by polymer and the transverse modulus of elasticity is no more than 10% of the axial modulus. Most designs of termination are, consequently, relatively long –

4 or 5 times the core diameter is a usual length of contact surface – and either a conical taper, to achieve high radial pressure, or compression jointing, is needed.

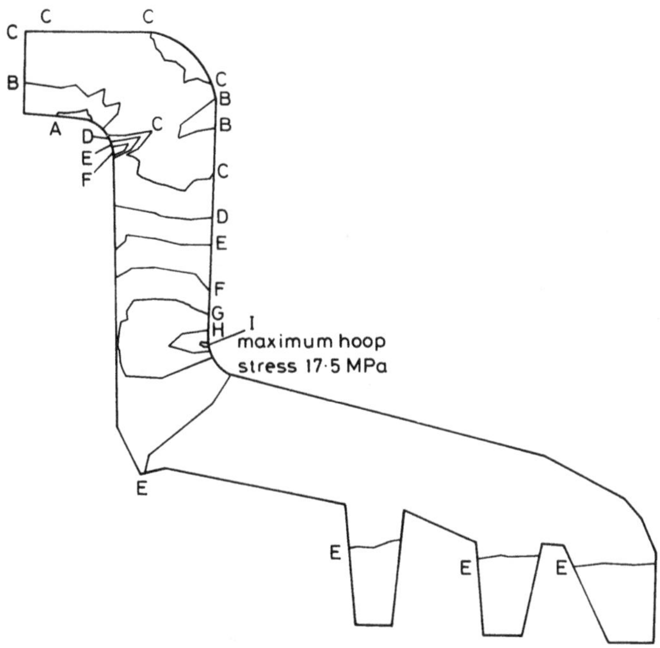

Fig. 7.7 *Finite-element analysis of disc insulator (after Reference 89)*
Calculated distribution of hoop stress in porcelain under 80 kN load on pin
Contour Keys:

	MPa
A	−16·0
B	−11·0
C	−8·0
D	−4·0
E	0
F	4·0
G	8·0
H	12·0
I	16·0

Some of the many coupling systems which are in widespread use are illustrated (Fig. 7.10). It is seen that either the end fitting or the rod, or even both, may be tapered. To taper the rod, either a profile may be machined at the end or the rod split axially, with some form of wedge introduced to spread the section. Compression jointing[84] involves a series of parallel swaging operations which impart a polygonal form to the exterior of the fitting. In this case, a progressive transfer of load is sometimes sought by decreasing the swaging

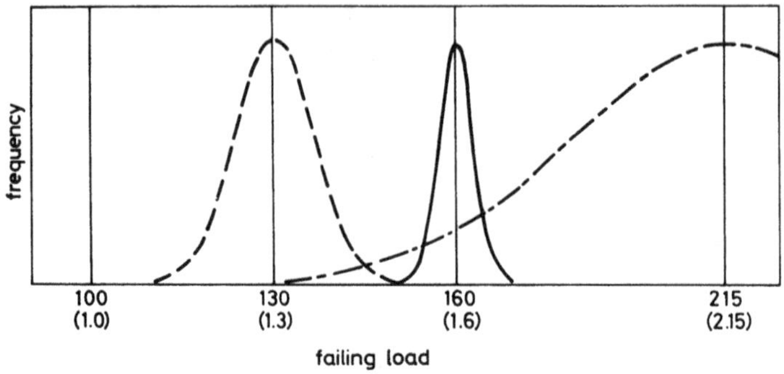

Fig. 7.8 *Variations in failure loads of metalwork and porcelain (after Reference 89)*
Distributions of failing load for cap, pin and porcelain, in 100 kN disc insulator
Safety factors above rating are given in brackets
—— pin
— — cap
—·— porcelain

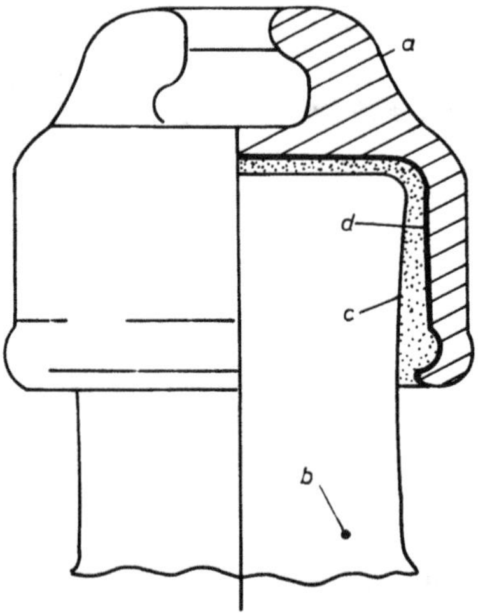

Fig. 7.9 *Terminal fitting for longrod*
a Metal cap
b Porcelain
c Conical face between porcelain and cement or alloy
d Bitumen coasting (optional)

pressure, and therefore the radially inward strain, progressively towards the fitting's mouth from its cap. The coupling area, which is in shear, is sometimes increased by machining flutes into the rod.

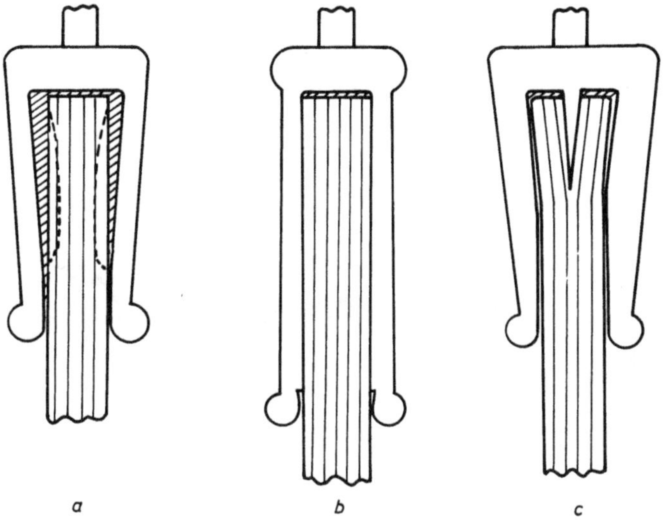

Fig. 7.10 *Polymeric-insulator end-coupling systems*
a Tapered fitting, tapered rod, together or separate
b Parallel rod, fitting swaged or cemented
c Internal wedge (see Fig. 7.11)

Analyses of performance of some of these systems, both theoretical and experimental[90,91], have been published. The principal points are as follows:

- Machining of the core is generally undesirable. Whether a taper is cut or flutes there is loss of effective cross section, since some of the fibres are interrupted.
- Splitting of the rod, however many sectors are produced, does not give a good fit to the inner conical face. Furthermore, at least one axial crack is made within the fitting which favours both mechanical and electrical weakness, especially if it accidentally propagates outside the mouth of the fitting (Fig. 7.11).
- For a rod of given quality best results, in terms of failing load, are obtained by using plain unmachined rod, well bonded to the correct thermoset, in a conical metal fitting face, the surface of which is lubricated. For equal strengths of the rod under shear and radial stresses a cone angle of about 5° is best.
- With most of the conical systems a proof-loading operation is needed in which the conical face slips and the core is locked to the thermoset.

Because of the anisotropic expansibility of pultruded RBGF, close to those of metals axially but much higher radially, reduced adhesion between metal and core must occur on cooling. It would be expected that the conical systems might accept relative contraction better than swaged systems, since additional slip should occur and the grip should be maintained with cones, whereas no such compensating process should exist where the load-transfer interface is cylindrical. Some published work does not support this hypothesis[90].

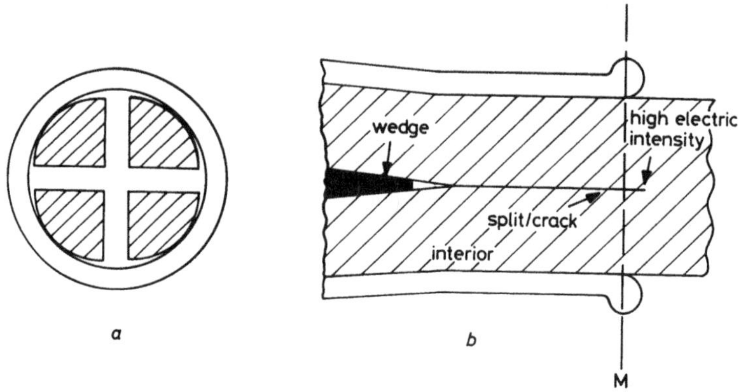

Fig. 7.11 *Objections to internal wedge*
a Even quadruple split gives misfit to metal fitting (exaggerated view)
b Any crack or split outwards from plane M is subject to large electric intensity

The materials used for fibrous core fittings include conventional galvanised iron, copper-based alloys especially for traction systems, aluminium and its alloys. For swaged systems only ductile metals are applicable. Because of the relatively small size of fitting in comparison with those needed for ceramic insulators, as well as the high mechanical stresses and need for good internal surface finish, expensive processes like investment casting are justifiable in some cases.

7.3 Effects of transition from metal to insulation

The performances of all types of insulator are largely dominated by effects which arise from the transition, at the terminals, from metal to insulating material. These effects include mechanical and electrical concentrations of stress, movements caused by differences in thermal expansibility and production of aggressive substances by electrolysis of surface contaminants.

Insofar as mechanical stresses are concerned, it is instructive to compare an insulator with the test pieces which are used to measure tensile strength of materials (Fig. 7.12). Even though the test piece is of a single material, it is necessary to make a gentle transition from the narrow neck in which failure is to occur, and where the tensile stress may be taken as uniform and calculable, to the broader part which carries the clamps or jaws of the pulling machine. No real insulator uses such a gradual transition. In virtually all types there is a step, at the mouth of the fitting, which would introduce a local shear-stress concentration, even if the fitting and insulator were of identical composition, simply on geometric grounds. In fact, there are usually large differences in elastic moduli and thermal expansibility between fitting and insulator which enhance the geometric effect.

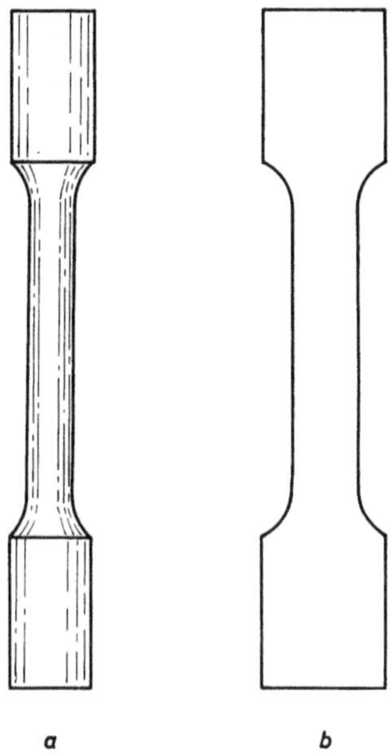

Fig. 7.12 *Standard pieces for tensile testing*
a Cylindrical
b Sheet
Terminal areas >2 × tested areas
Transition radii > tested widths

Mier-Maza *et al.*[90] show the sizes of these concentrations for different designs of fibrous core fitting; they are large even in the absence of thermal effects and of bending loads, which are commonly superimposed on pure tension for many designs of post and rod in actual service.

Differential thermal expansion gives rise to tensile stresses in the insulating material wherever insulation and metal are cemented together over substantial lengths, for which reason controlled slippage is designed into terminations for longrods and disc insulators. Again, the stress concentration arises at the mouth of the metal fitting.

The electrical-stress concentrations arise from different causes and are not all equally objectionable. When the transition is clean and dry the electrostatic

Table 7.2 *Cements and coatings for insulator fittings*

Property	Portland	Aluminous	Remarks
Compressive strength (MPa)	45	75	These values for mix with 25% silica sand: strength and modulus are lower for pure cement paste
Young's modulus (GPa)	32	40	
Expansibility ($\times 10^{-6}/°K$)	9·2	10	These values for range $-30°C$ to $+40°C$
Electrical resistivity (max., dry) (Ω cm)	10^{12}	10^{12}	Fully wet concretes have resistivities as low as 10^3 Ω cm; permittivity and tan δ are also variable with humidity

Coatings for cap interiors and pin exteriors are generally bitumens (asphalts) or polymers such as polyurethanes, nylons and polyisobutylene-blends Thicknesses are of order 10 μm to 100 μm: uncoated metalwork causes reductions in tensile failure load up to 50%. Low-temperature strength and ability to survive load cycling both depend strongly on presence and nature of coatings.

distribution of stress is determined by the geometry and permittivities, respectively infinite, from about 3 to 7, and unity for metal, insulation and air. In this condition the consequence of discharges is radio interference, but the option is usually open of a cure by external stress grading. In moist or lightly polluted conditions the usual effect is production of a dry band close to the metalwork, for reasons described later (Chapter 11). Discharges span this band and may be of a severity sufficient to damage both polymeric housings and glass discs. In severe pollution, when pulses of leakage current having amplitudes of ampere order are flowing through the surface layer, serious erosion of polymers and glass is likely at the transition because of the energy of the arcs.

In the polluted condition, electrochemical products of electrolysis in the wet surface layer are generated, in quantities related to the leaked charge, at the boundary between metal and insulation. These are not transitory and will continue to attack both metal and insulation until washed off or otherwise removed.

It is seen that, especially for fibrous-cored insulators in which the rate of loss of strength depends strongly on the mechanical stress, the temperature and the presence of chemical aggressors, the transition region is a likely initiation site for failure; this is observed in practice[49].

Chapter 8

Finite insulator life: limiting processes

8.1 Catastrophic and gradual attack

An insulator comes to the end of its working life either when it fails mechanically, flashes over at unacceptably high frequency or gives evidence of deterioration to a condition likely to lower its factor of safety in service. All insulators are affected to some extent by impact, cycling both thermal and mechanical, ablation from weathering and electrothermal causes, flexure and torsion, ionic motion, corrosion and cement growth. There are, however, strong differences between ceramic and polymeric insulators, as classes. In general, a ceramic insulator will be vulnerable to impact damage, since its dielectric is a brittle material, and to processes which cause concentrations of tensile or shear stress. If porcelain, it will be near absolutely resistant to chemical and discharge attacks. On the other hand, a polymeric insulator will resist impacts, in handling and in service, but will be damaged by chemical and even weathering processes as well as by discharges. Glass occupies an intermediate position in being largely immune against atmospheric chemicals and pollutants, with the exception of some halides, but suffering much more than porcelain from discharge attack: it is of all types most susceptible to deliberate vandalism by shooters.

Catastrophic damage to porcelain includes: loss of part of the profile, with consequent reduction in resistance to flashover; cracking of the head of a disc insulator (which introduces the possibility of a head burst on subsequent passage of fault current), either by cement growth, pin corrosion, impact or lightning puncture; loss of water seal in a bushing shell or cable sealing end by radial cracking of the cylindrical part; weakening in cantilever of a post insulator, again caused by cracking of porcelain in an analogous way to the types of failure in discs.

For very few of these events is immediate detection easy. Especially with cracked housings, progessive water invasion has insidious and expensive consequences, while loss of cantilever strength in a substation post may only manifest itself by collapse of a busbar under fault-current mechanical load. In recent times, therefore, attention has been given to reducing impact damage by the use of resilient protective fittings or sleeves and to the replacement of types which are vulnerable to corrosion or cement growth by solid posts.

Although deterioration by development of microcracks must occur internally, surface attack on porcelain is negligible in shortening an insulator's life: fifty-year old discs have been found discoloured but sound. The fittings, however, are often seriously decayed, with caps bare of zinc and rusted. Very old pins, especially from wet areas, have lost up to 30% of their strength by necking near the cement surface, where corrosion has been most severe. Corrosive effects are much more severe under DC than AC (see Section 15.4).

In considering catastrophic attacks on polymeric insulators the most dangerous failures have been mechanical, from loss of strength in the core. These have arisen from: pull out, i.e. loss of bond between the end of the core and the metal fitting; from rough handling during transport or erection which has weakened the core near one terminal; and from brittle fracture of the fibrous material[59–62]. It is sometimes overlooked that vandal attack on polymeric insulators which leads to loss of protection of the core, for example by cuts or cracks arising from bullets or lead shot, may be especially dangerous as not only undetectable by inspection but also as leading eventually to collapse of the core under load.

Gradual attack on polymeric insulators includes: loss of surface-water repellency by physical and chemical degradation, leading to grossly reduced flashover voltages in a given severity of pollution; erosion of the housing both distributed and localised; loss of mechanical strength in the housing leading to cracks and relaxation of grip on the core; undermining of skirts by erosion causing an avalanche process with increasing leakage current; development of a conductive path at the housing/core interface. It is even more difficult to detect any of these effects than those on ceramic types: measures like monitoring of visually evident discharges, detection of thermal changes by infra-red telescopy, measurement of interelectrode voltage or resistance, all of which have given some success with some types of ceramic insulator, are generally inapplicable to polymerics[35].

8.2 Impact testing and vandal resistance

Some indication of the ability of an insulator to resist impact damage in handling, and from low-velocity missiles like stones and bottles, is obtainable by the pendulum test[92,93]. A bob is allowed to swing down and strike the insulator, thus measuring the kinetic energy which is required for fracture of an exposed part, such as a skirt or shed. On most kinds of ceramic or glass insulator values of only a few Joules are obtained – a high-strength alumina post may show 3·5 J, for example – which are well below the energies of typical missiles.

Practical tests have been made on the efficacy of a layer of resilient polymer as a means of absorbing impact energy. A few millimetres of elastomer have been shown to be surprisingly effective in preventing fracture of a porcelain post by shotgun fire[76] at a range as short as 6 m. These tests showed incidentally that the shot reached the porcelain core and would, accordingly, if directed against a fibrous-core insulator, have produced the very type of insidious damage mentioned above.

Anti-vandal polymeric protectors for porcelain insulators are currently available in different forms, which give varying areas of shielding (Fig. 17.5). In the Hybrid insulator the whole surface is protected, with the trivial exception of the refractory bands. The vandal shield, invented by Salmon[94], who first drew attention to the severity of the vandal problem, covers most of the surface but only from some angles of attack. Creepage extenders and rim buffers protect only the edges of the skirts and are therefore ineffective against shot or bullets. All protectors have some effect on the electrical performance of the insulator to which they are applied; this ranges from enhancement of flashover resistance, by Hybrid sleeve and creepage extender, to interference with normal rain washing, and hence some increase in discharge activity, for the vandal shields.

Surprisingly, in the case of glass discs for which anti-vandal protection would appear most needed, no specific remedy appears to have been developed at the time of writing.

8.3 Damage by cycling

Cyclic loading is a known cause of material failure since it both promotes growth of micro-cracks directly and allows ingress of water to all kinds of surface flaw. In cap-and-pin disc insulators care is taken to accommodate cyclic changes by mechanical design and use of what amounts to high-pressure lubricants on vulnerable interfaces. In the case of thermal cycling the relative expansibilities of the metal fittings, cement and dielectric determine the sizes of stress which are generated, not only by temperature excursions from full sunshine to clear night sky, which may exceed 80°C, but also by heat generation under passage of fault-current arcs.

The thermal mismatch between porcelain and both metals and cements of the types used in manufacture is large; for glass it is much less (Tables 2.2 and 7.1). Remembering that it is only in tensile or shear stresses that ceramic dielectrics are weak, it is seen that thermal cycling is a probable cause of insulator death only where such stresses can be generated. In practice, this confines thermal cracking to discs, pedestal posts having internal metalwork, and line posts loaded in cantilever. For longrods with ceramic in tension, additional stresses, either tensile or shear, are generated by the metal fittings. Lubricated conical interfaces or compliant metal alloys, or both, are incorporated to minimise the effects of these.

The case of the multiple-cone insulator is a special one. Cracks have been found in old busbar supports, the origin of which is not clear. Thermal mismatch between the porcelain cones and the cement is one possibility, but another is cement growth.

8.4 Cement growth and corrosion

Although sulphur-based and metal-based fusible fillers as well as aluminous 'Ciment Fondu' are used in some assemblies, the predominant cement in insulator manufacture is Portland, either neat or as mortar with added mineral powders. The outstanding difference between a true cement like Portland or 'Fondu', and other types of filler is that it embodies water and has an electrical conductivity, based on ionic motion, which varies very widely as water migrates in or out of it, even when it is both fully cured and effectively free of large pores.

Portland cement is an impure mixture of calcium silicates and calcium aluminate, with excess calcium oxide. Gypsum, i.e. calcium sulphate, is present and iron, as a ferrite. On addition of water the silicates react to form a gel-like solid, mainly of calcium silicate hydrate and similar in structure to a clay mineral. The aluminate, ferrite, lime and gypsum react with water to form other gels and crystals. Although curing, in steam, water or both, is nearly universal practice, some unreacted components are often left, and subsequent changes, in composition as well as crystal structure, occur, sometimes over periods of years.

Atmospheric contaminants, especially sea-salt, road-salt and certain sulphates, are known both to attack Portland cement itself and, if allowed to permeate or migrate onto any buried metalwork, to corrode it or react with its surface galvanising zinc. The latter effect is much aggravated by ionic migration under applied electric fields. Where this is unidirectional, for example in the case of disc insulators used for DC transmission where the buried pin is the positive electrode, internal corrosion has caused cracking of the ceramic head in a matter of months. Under alternating stress the rate is reduced by a factor of some hundreds or more, but the effect still persists. Heavily loaded and salt-contaminated strings of insulators in the UK have demonstrated close correlation between locations of maximum cracking rate and insulator-unit voltage maxima; there seems litle doubt that metallic growth by corrosion is a major cause of insulator failure.

On cement growth the picture is far from clear. Cases have occurred, as with multiple-cone posts, of cracking which could not have been caused by metallic corrosion, since the only materials in play were porcelain and cement. Cherney[95] pointed out that many cements used in North America contract rather than swell with age, and that growth can be produced only by highly unrealistic thermal cycling. It has been suggested that the gypsum, which is added to moderate the high reaction rate of the tricalcium silicate in Portland cement, may cause swelling; other sulphates are certainly known to be undesirable. The structures of hydrated cements are, however, so complex and the possibilities of long-term reactions so many that diagnosis of alleged growth is difficult. One possibility is effects of displacement current: the loss angle of cement varies widely with humidity, as therefore must the internal heating and expansion caused by this current.

The extent of the cracked-insulator problem has been increasingly realised since the 1970s, with the recognition that mechanical factors of safety have declined. This was first published in Canada, where between 30% and 50% of samples removed from lines of various ages were found defective by test, and later in England[15,96]. There, somewhat smaller defect rates than the Canadian appeared, but still of concern as lying between 10% and 15%, in this case on important 400 kV lines. In Canada many of the failures on recently made insulators were attributed[95] to excess gypsum in the cement, while in the UK the cause was claimed to be defective protection of metal fittings. There seems no common feature of duty in Canada and the UK for insulators. Especially on the 735 kV lines in Quebec there is large electric stress, since relatively few units per string are used, but the levels of salt pollution are very low; on the English 400 kV lines there is low unit stress with very high salinity levels. At least these facts are consistent with the view that all types of corrosive-generated cracking require higher than normal passage of leakage charge through the cement and into the buried metal.

8.5 Loss of electrical performance

With the exception of resistive glazed porcelain, dealt with in Chapter 14, only polymers and glass suffer significant loss of electrical performance with time in service.

Loss of electrical performance arises from unacceptable increases in leakage current, with given pollution level and surface electric gradient. These increases are caused: by thickening in the layer itself, because of surface roughening or chemical bonding between surface and pollutant; by gross damage in the form of channels or tracks; and by actual loss of creepage path arising from undermining of skirts or loss of complete units.

Very light frosting of glaze is sometimes seen on porcelain after heavy and repeated electrical activity, and a short-term enhancement of surface adhesion also occurs then, the so-called conditioning which is seen during artificial salt-fog testing. Fault-current arcs sometimes cause spalling of glaze and staining with electrode material. Since, however, all electrical porcelain is fully vitrified there is no question, as is sometimes supposed, of pollutant soaking into a porous interior. Even where blown desert dust or sand[97] abrades the glaze to expose the white body the increase in surface roughness is trivial, although there is loss of mechanical factor of safety in all cases of glaze damage.

With glass insulators in heavy salt pollution, and especially where there is ancillary clay, cement or other inert absorbent material, frosting of the central regions, where densities of leakage current are high, appears quite early after energisation. This is undoubtedly an effect of leakage current: it is not seen on electrically by-passed insulators used at the earthed end for instrumentation in testing stations. Later stages of damage by leakage current include treeing, in channels of the order of 1 mm deep, and cutting of skirts or sheds, especially that closest to the pin. Damaging flow of current caused by heating can occur in insulator glass where surfaces are blanketed by deposits. The usual consequence of a progressive increase in leakage current is the shattering of a unit which, unless rapidly detected and remedied, causes a large upward step in leakage current, and consequently also in probabilities of further shatterings and of flashover.

The saving grace with glass insulators is that shattered units are immediately visible, by inspection from ground or helicopter. No comparable indicator of approaching trouble exists either with porcelain or polymer, for both of which detection of deterioration is difficult and expensive.

It is with polymeric types, and especially those dependent on external materials which are heavily filled and vulnerable to weathering, that decline of electrical performance is most marked (Section 6.5). Where both the surface is crazed and filler is exposed, very high values of surface electrical conductivity in pollution are obtained. Temporary relief can be had by treatment with hydrophobic silicone oils. Undermined skirts and concealed tracks along fibrous cores both lead to sharp declines in flashover voltage. Flashovers which occur internally may cause explosion and mechanical failure. In practical terms, none of these deteriorations is easily detectable, even by close inspection: unfortunately, because of their very low end-to-end capacitances, polymeric insulators which are on the point of failure do not give warning, in most cases, by the generation of electromagnetic interference at enhanced levels. A defective insulator of conventional type can often be detected in this way, and even located accurately when it emits ultrasound or high-frequency radio waves.

An unexpected quality of many polymeric insulators has proved to be a high resistance to damage by fault-current arcs. The metalwork is usually much less massive than on conventional insulators, giving a somewhat enhanced risk of fatal damage by arcs which are allowed to root and cause significant melting, and there is in principle also the possibility that swaged-on fittings will relax their grip. However, as far as the dielectric is concerned the survival rate is high: even heavily sooted insulators, with their surfaces burned and ablated by long-duration arcs at kiloampere level, have purged their surfaces and given normal counts of leakage-current surge after short exposure to weather and pollution. Housing polymers which contain active fillers have behaved especially well in this respect[98].

Damage to conventional insulators by arcs may include shattering of glass discs, cracking of housings and mechanical failure of longrods. Modern circuit-breaker designs result in fault durations of very few cycles, much reducing the risk of thermal damage. Arcing horns are, however, still conventionally used outside some parts of the USA to protect insulators and conductors from damage: some ingenious designs use the interaction of the arc with the magnetic field of the fault current to move the arc roots in desired directions.

Chapter 9

Aesthetics of insulators

9.1 Acceptability of transmission lines

Public opinion is now so powerful a force in many parts of the USA, Europe, and especially Great Britain, that the prospective visual appearance of a power line may be crucial, in deciding whether or not its construction will be allowed. For example, in some parts of England and Wales which are of outstanding natural beauty the construction of orthodox steel-towered lines, using vertical suspension strings, is not permitted. Only those constructions which are inconspicuous, both in height of tower and choice of insulator, are accepted. Similar constraints have led, in North America especially, to radical proposals for compaction of power lines and to strategic analyses of the consequences of replacing 'power corridors', in which several lines run nearly parallel, by one or two circuits at higher voltage.

Underground cables have been quite widely used, especially in Great Britain, not only in urban areas but also in places where towers would be intolerable. The costs of undergrounding, both direct in terms of money and indirect, arising from incidental damage to the environment, are very high: replacement of one of the CEGB's standard 400 kV double-circuit lines by cable has been estimated to cost, in terms of capital, 20 times more. In addition, the required civil works, taking account of the need to remove ohmic heat by water-cooling lines, are comparable with the construction of a four-lane motor road. If, therefore, the alternative is undergrounding, quite high costs are justifiable in a 'beautiful' line which will be acceptable by the environmental protectors.

Insulators will play two roles in beautification or compaction, respectively in being themselves of pleasing appearance and in permitting special designs of supporting structure which are lower than conventional, less conspicuous, or themselves more analogous to natural forms. An example is the Trident line (Fig. 9.1), which uses special insulators to achieve, at 132 kV, a near halving of structural height, together with excellent compatibility with trees in visual terms.

Fig. 9.1 *'Trident' line: 132 kV – 100 MW rating*
Polymeric line posts, central one inclined along line, avoid cascade failure. Structure merges into wooded landscape, is much lower than equivalent steel tower

9.2 The inconspicuous insulator

The quantities which decide how conspicuous an insulator will be, when installed on a line – little can or need be done about the appearance of insulators in substations – include its colour, silhouette area, shape, orientation and whether it is single or part of a duplex set or cluster.

The two common colours for porcelain insulators are chocolate brown and 'Munsell gray', the latter being a pale blue which is intended to match a lightly overcast sky. There are technical disadvantages attached to the requirement for a glaze to be coloured (Sections 3.4); only about half of the proportion of 'stainer' which is needed to give the chocolate colour (7–9%, mostly iron oxide) is used for the Munsell gray, usually oxides of nickel and chromium. Many traction insulators use transparent glazes: the visible body gives an ivory colour to the finished product. For transmission insulators, where a compression glaze is used in some parts and a tinted one elsewhere, it might be thought sensible to use only the former, which again is nearly colourless. However, these compression glazes tend to have matt surfaces and consequently to be less efficiently rain-washed than normal, glossy ones.

Since insulators are seen, almost always, against a background which is bright, their surface-reflected colour is not dominant. It is generally difficult to distinguish different coloured porcelains, or even porcelain from glass, at least for transmission lines where the insulators are seen from ranges of many metres. Silhouette area is a different matter: there is a very great advantage, when seeking the inconspicuous insulator, in going from cap-and-pin to longrod, and from longrod to polymeric, because of the reductions in overall diameter: typically, for equal mechanical rating, the diameters will be as 4:2:1. This advantage can be achieved only in some ranges of voltage and mechanical loading, however, insofar as porcelain longrods are concerned, because their cores fail at 160 kN. For greater loads than this, i.e. for the requirements of transmission at voltages above 220 kV, duplex sets or clusters are needed: the benefit of the slim longrod is then lost, since clusters are much more obtrusive than single strings or rods (Fig. 9.2).

The advantage of using polymeric insulators increases rapidly with voltage, in respect of silhouette area, and also with the number of subconductors, on which the wind loading, and therefore the required tensile strength of the insulators, in terminal and angle positions depends. The use of quadruple strings in tension positions, as in the CEGB's 400 kV standard construction, gives a highly conspicuous structure. Substitution of high-strength duplex sets results in modest improvement, but polymerics, if they were available with sufficient reliability, would transform the appearance totally (Fig. 9.3).

For the UHV or million-volt transmission range again the polymeric insulator would be uniquely inconspicuous. Some projected and actual arrangements (Fig. 9.4) show how marked is the improvement when insulators are used

Fig. 9.2 *Obtrusive clusters of insulators*
Line designed to VDE Specifications (Photograph by kind permission of Balfour Beatty Transmission Division)

which are not much different in apparent area from the conductor bundles and structural members.

9.3 Insulators as determinants of tower height: Compaction

In the simplest case of a suspension tower for double circuits, the length of the insulator string has a direct effect on the height of the tower and the length of

Fig. 9.3 *Inconspicuous polymeric insulators*
Slim, pultruded-core polymeric longrods (far circuit) are much less visible than duplex sets (near circuit, top phases). Laminated-core polymeric longrod (bottom phase) is of higher strength for larger conductor bundles

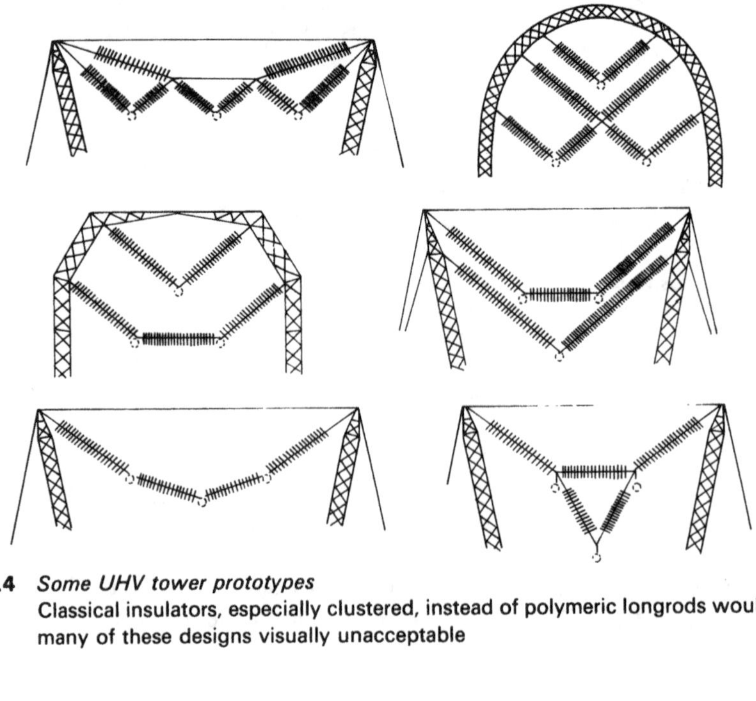

Fig. 9.4 *Some UHV tower prototypes*
Classical insulators, especially clustered, instead of polymeric longrods would make many of these designs visually unacceptable

the crossarms. Since the cantilever load at the foot of the tower depends strongly on the height, the cross-sections of the structural steelwork, and hence the opacity of the tower, must also rise as the tower is made higher. Fig. 9.5 shows how shortening insulator strings confers benefits, not only from lower towers but also from shorter crossarms and reduced sections of steel, in cheapening construction and in making the structure less conspicuous.

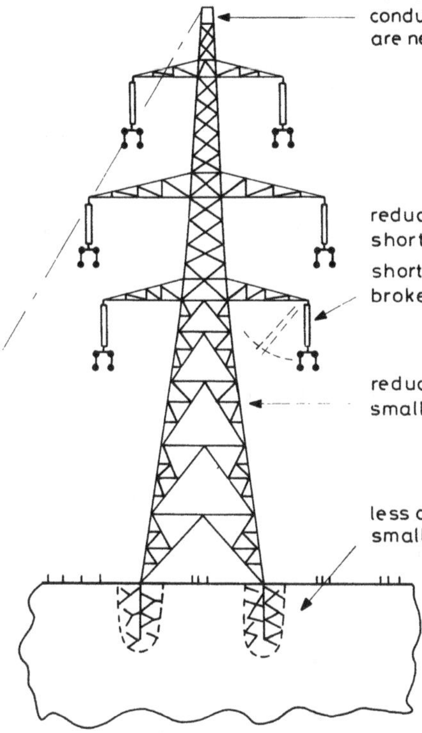

Fig. 9.5 *Benefits from shortening insulator strings*

It is already possible to reduce the heights of towers by using either inclined or horizontal strings in place of vertical ones: this has been common practice, for example around airports, in Europe for more than ten years. Any benefit, visually, is usually lost because of the greater obtrusiveness of the replacement insulator clusters (Fig. 9.6), although once again the availability of slim, strong polymeric insulators would transform the situation. Very low structures also result from insulating crossarms, disposed in Π formation (Fig. 9.7). A rather unpleasantly squat appearance results in the support structure, and the arrangement is open to criticism for employing phase-to-phase insulators, on which flashover causes difficulties for the protective circuitry.

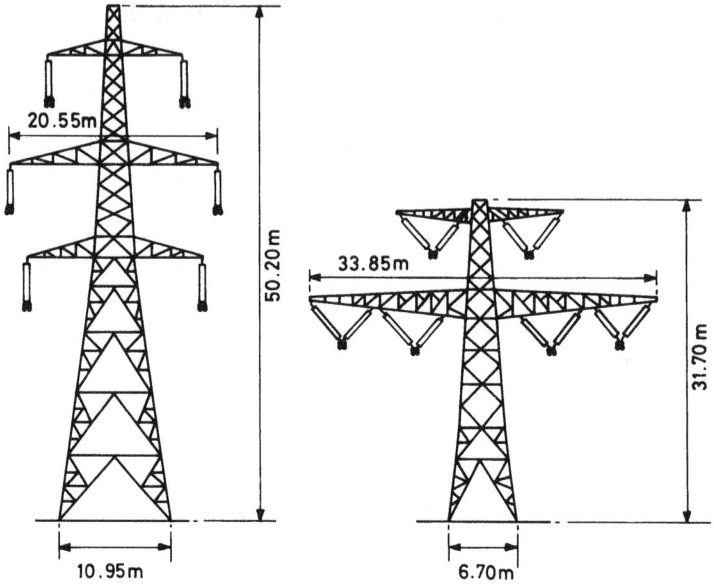

Fig. 9.6 *Standard and 'low height' towers*
Both designs support 4 × 400 mm conductors per phase: 400 kv system (CEGB, UK, designs)

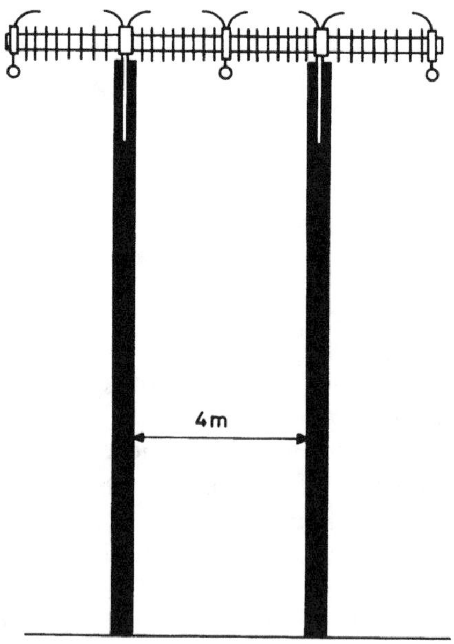

Fig. 9.7 *Insulating crossarms in Π-formation*
Minimum-height construction for 132 kV system

Insulators may also contribute to compaction when used to prevent infringement of flashover clearance between closely spaced conductors (Fig. 9.8). Insulators which are light in weight are highly desirable for such applications, again pointing towards polymeric or special longrods since the mechanical functions include actions both as strut and tie for which discs are unacceptable.

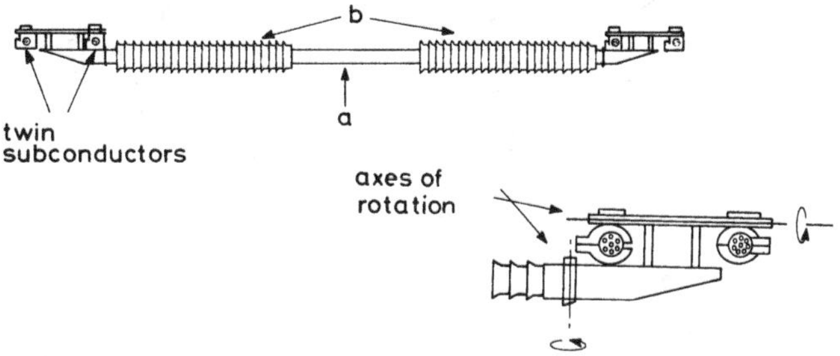

Fig. 9.8 *Interphase separators*
Sections (*b*) are insulating; (*a*) may be metal.
Gimbal couplings are used with ceramic elements, to minimise bending loads: unnecessary with polymerics

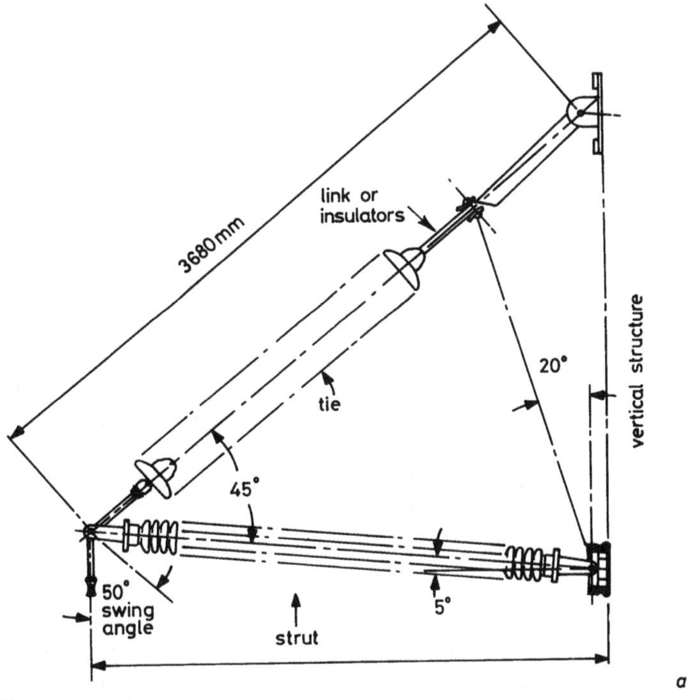

a

Fig. 9.9 *Two types of compact support*
a Polymeric insulating tie or string of discs combined with porcelain post (Lapp Insulator Co.)
For 220 kV, assuming freedom to rotate, mechanical ratings are: tie = 100 kN; strut = 120 kN
b (Overleaf) 'Oscillating' polymeric line-post (Electricité de France)

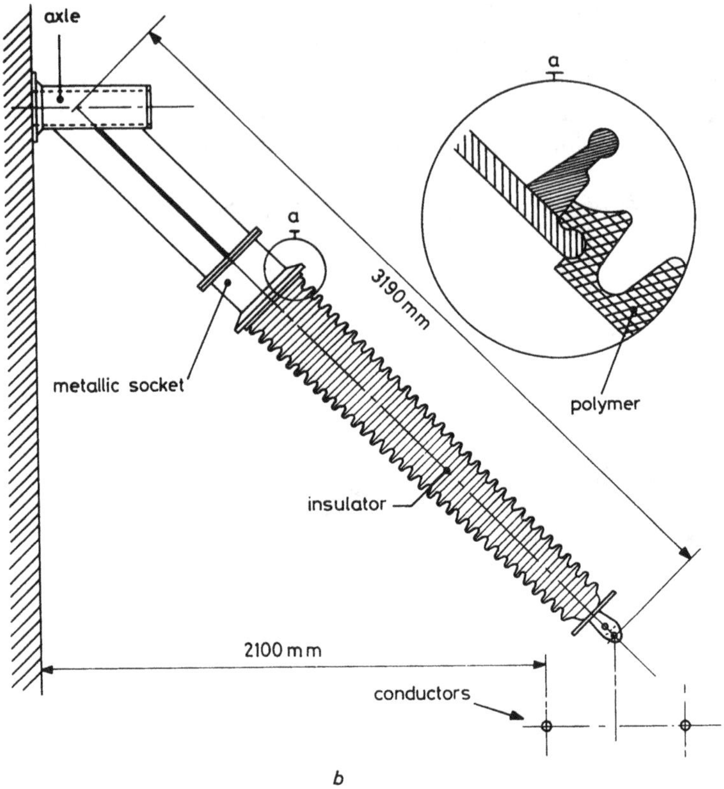

Fig. 9.9 *(continued)*

Numerous combinations of separately insulating tie and strut are in use, or under consideration, as means of compaction (Fig. 9.9). The broken-wire condition, where failure of a conductor occurs and places large asymmetrical loads on the structure, is difficult to meet with many of the compact supports; shear pins or hinging arrangements need to be incorporated, if cascade failures are not to occur along several spans of line[99,100].

9.4 Unorthodox systems

Several unusual systems are in use or under development, the object of which is to improve the appearance of transmission lines, especially those which run close to, or even inside, cities. These include curved crossarms and vertical structures based on polymer concrete (Fig. 9.10). A reasonable aim must be to combine the support and insulating functions into a single structure. Although this can already be done for very low voltages by the use of lattice sections based

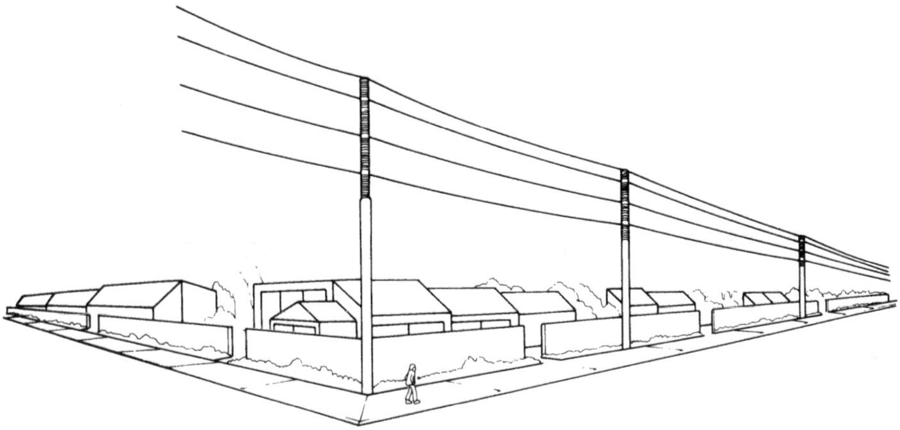

Fig. 9.10 *Proposed 'Polysil' posts (after EPRI Journal, May 1983)*

on fibrous composites, extrapolation to transmission voltages calls for materials better than those currently available, to deal with combined electrical and mechanical stresses in pollution. Some speculations are made in Chapter 17.

Chapter 10

Physics of contamination

10.1 Electrically significant deposits

The terms 'contamination' and 'pollution' have special meanings when applied to the condition of insulators. An insulator so heavily polluted by marine deposits that it flashes over immediately on energisation may appear to be perfectly clean, even on close inspection. On the other hand, one which is black with industrial soot, or which has some of its surfaces caked with cement, may have an electrical performance indistinguishable from that of a freshly installed counterpart.

The reason for this apparent paradox is that values of surface electrical conductivity which are sufficient to cause flashover are quite trifling in absolute terms. They are readily achieved by the presence of soluble electrolytes, such as common salt or industrial acids, at densities of some 0.1 mg/cm^2, provided water is available to dissolve them. They are not readily achieved by layers of carbon particles, which make only intermittent point contacts with each other, or by aggregates of mineral dusts which are free of ionic components (although combinations of such aggregates with soluble salts, giving a 'blotting paper' effect, have caused severe flashover problems in North Africa and in Cornwall, England).

The deposits which are of greatest significance, in the performance of insulators, are therefore highly soluble electrolytes originating from the sea, from road-salt, from salt-flats and desert dusts, and from industries such as petrochemical and other acid generators; less dangerous, although locally important, are the above-mentioned aggregates, fly-ash from generating plant which burns pulverised coal, and industrial fumes. The latter are in any case under attack as environmental nuisances which are avoidable. Both the soluble ionic and inert layers require water before they can act: fog, dew and drizzle are thus also highly significant deposits.

Pollutants which remain electrically conductive even in the absence of water include carbon, some metallic oxides and metals in the form of dusts or powders. Flashovers directly caused by these are rare; however, in the case of railway insulators, considerable contamination may be caused by oxides of iron from the wheel-brakes, or by carbon or copper ablated from conductors or pantographs. Such contaminants may be reactive with polymeric insulators and with insulator greases, and have caused severe damage in association with other pollutants: special tests have been devised to accommodate conductive pollutants[80,81,101].

A special case is that of the very fine silica dust used in the manufacture of some protective silicone pastes which are applied to insulators. Destruction of the silicone, by discharge and weathering, allows the silica to adsorb water in such quantities as to cause local damage by Joule heating[102].

Consideration is now given to the two classes of process which determine the equilibrium state of contamination, respectively deposition of new material and purging or washing away of old.

10.2 Contaminating processes

The principal processes which transport material onto the surfaces of insulators are gravitational forces, electrostatic attraction of electrically charged particles, dielectrophoretic migration of high-permittivity particles into regions of large electric-field divergence, evaporation of solutions or suspensions and aerodynamic catch. The last is entirely predominant in importance[103].

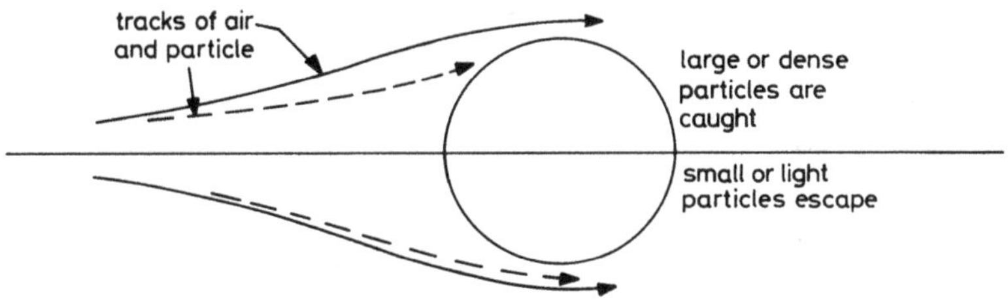

Fig. 10.1 *Dependence of catch on nature of particle*
Particles of small radius or density are held in the diverted flow by viscous forces

When air containing suspended particles flows towards an insulator, the efficiency with which the insulator catches particles depends on the shape of the insulator, on the size and density of the particles and on the speed of the flow. The insulator forces the flow to divide, leaving a stagnation point where the air is at rest. The flow changes direction away from the stagnation point, but the suspended particles, having densities greater than that of air, are unable to follow the flow accurately and pursue paths of lesser curvature: this motion relative to the surrounding air is resisted by the viscous forces on the particle, however, and is small for low particle diameters and densities (Fig. 10.1).

The simplest case of particle catch is thus the deposition of a relatively large, dense droplet or granule at a point of stagnation from which there is no force to remove it. Much more common and important, however, are the effects of rotating-flow or vortex generation, arising from the disturbance to the air flow introduced by the insulator. Vortices are produced at the sides of the insulator, which affect other insulators in its wake, and also within the underside structure of the insulator itself, especially by deep skirts or sheds. Rotating flow of this nature gives rise to cyclone action: a given population of particles will be trapped in a rotating volume for many cycles, and the time for migration to the wall of the insulator, against the viscous forces, will be prolonged. Many quite small and low-density particles will, in this way, be deposited and, moreover, actually within the convolutions (Fig. 10.2).

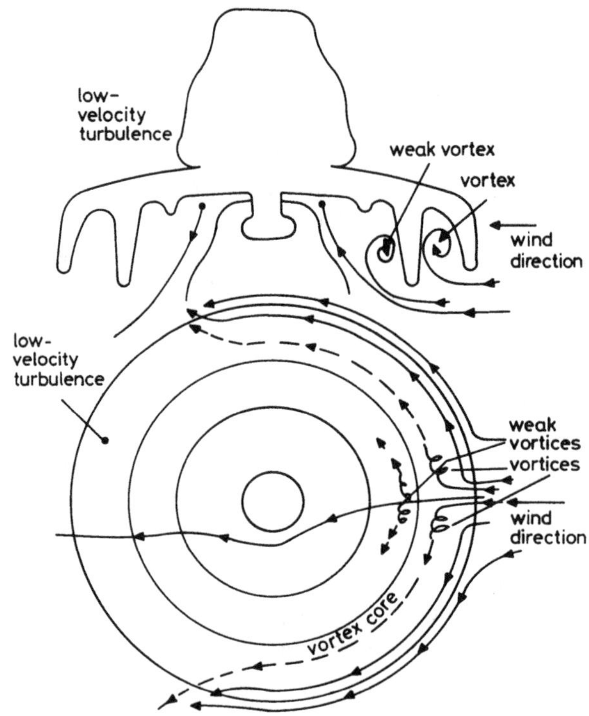

Fig. 10.2 *Observed flow over anti-fog disc (after Walshe[104])*
Smoke paths at 2·4 m/s show vortices between skirts

The possibility that a significant improvement in performance of insulators might be achievable by modifying shapes, in the direction of lesser interference with incident air flow, seems first to have been investigated by the author and B. F. Hampton[104]. Studies in wind tunnels were made on their behalf by D. E. J. Walshe using two types of disc insulator and a dummy biconvex aerofoil. Flow patterns were measured, using paraffin smoke and also titania in oleic acid as a surface indicator. Actual deposits were quantitatively compared, using talcum or suspensions of magnesia as artificial pollutants, injected into the flow upstream of the experimental pieces.

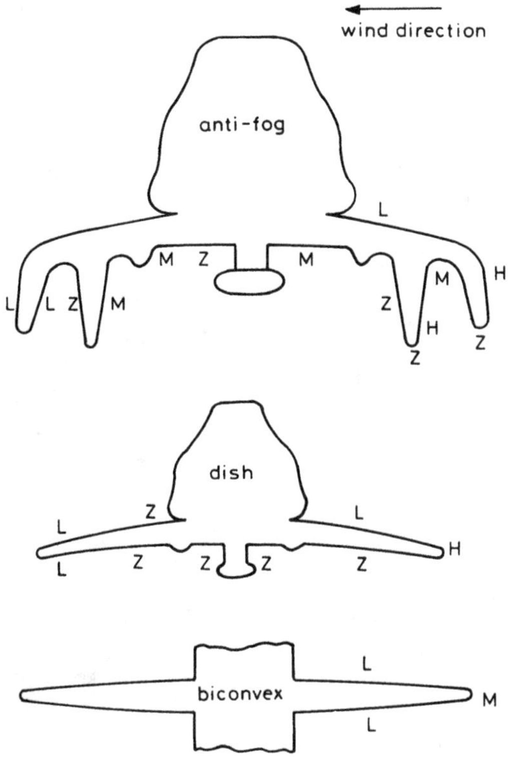

Fig. 10.3 *Variation of pollution catch with shape*
H: heavy
M: medium
L: light
Z: zero deposit density

Catch of pollution on underside (mg)		
at 9 m/s		at 1.5 m/s
630	anti-fog	150
390	dish	10
40	biconvex	negligible

The results (Fig. 10.3) showed that maintenance of high flow speed over the surfaces and elimination of vortices, especially those within convolutions, had had dramatic consequences for the catch of pollution. The quantities caught beneath the heavily convoluted discs were more than ten times higher than for the biconvex shapes, but density maxima were as 100:1 or higher, with strong concentrations of deposit associated with the edges of the skirts.

Once deposits have significant thickness they have the secondary effect of modifying the air flow, both by increasing the frictional drag and by causing subsidiary vortices. Heavily convoluted insulators may become clogged as a result of such cumulative catch, with disastrous effects on the electrical performance by the loss of effective creepage path (Fig. 10.4).

Fig. 10.4 *Clogging: anti-fog insulator from bulky-polluted tower* (photograph ack. CERL Leatherhead)

Effects of shape on aerodynamic deposition are less marked with different designs of large post or cylindrical insulator, since the flow is dominated by the central core. Nevertheless, improved flow has been claimed for designs where only slight dishing of skirts has been used, with close spacing and alternate large and small diameters of skirt (Fig. 10.5)[105].

We have considered the impact of particles as a cause of pollution, but we must also have regard to impacts which may remove deposits. Raindrops range from about 0·1 to 4·0 mm in diameter and follow trajectories, even in high winds, which are much different from the flow lines. They will hit and clean upper surfaces of insulators as well as cores and bluff edges of skirts, but will not penetrate into convolutions. Grains of sand, again of high density and diameters up to 0·1 mm, will similarly purge only outer surfaces. The so-called 'protected creepage' on many convoluted insulators is thus seen often to be more in the nature of 'protected dirt' (Fig. 10.6) (see, however, Section 6.6).

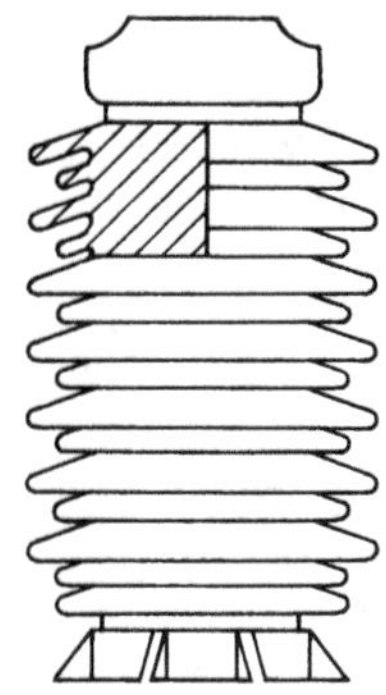

Fig. 10.5 *Desert-design post*
Close-spaced sheds, lightly dished, promote low catch of pollution
Thick central core perturbs air flow: design is subject to flashover under heavy wetting

The importance of aerodynamic effects in the pollution of insulators lies not only in their universality – every insulator, whatever its electrical condition, is subject to airborne contaminants – but also in their long range. Fine dusts containing salt and dry plankton, which are generated by breaking of ocean waves, may be carried for tens or hundreds of kilometres to contaminate inland power lines; visible deposits of matter blown from the Sahara are claimed to have been found in English car parks. Electric and magnetic fields act over relatively insignificant ranges.

Many airborne particles are electrically charged by triboelectric or frictional effects, and by attachment to ions generated from cosmic rays or industry. Such particles will have an electric component of force added to their gravitational and aerodynamic ones, and will be caught when they come within range of appropriately charged DC electrodes, but will remain free in alternating fields.

Contamination patterns are significantly different, for this reason, between insulators on DC and AC circuits. Where corona activity causes large local ion fluxes, intense dirt deposition occurs, again on DC only, by electrostatic precipitator action.

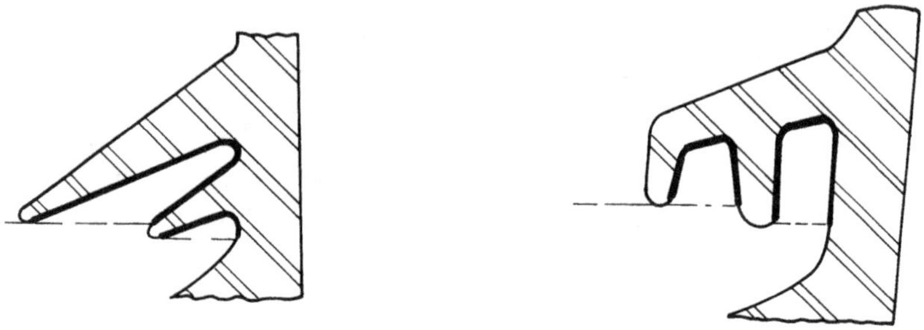

a *b*

Fig. 10.6 *'Protected' creepage*
Both shapes have 50–60% 'protected' creepage (*a*) accessible to washing (*b*) inaccessible; (*a*) has much better performance

The motion of particles having high permittivity into regions where the divergence of electric intensity is large, dielectrophoresis, is a very short-range effect, which, however, is polarity independent. The force depends on the volume, relative permittivity, but not the state of charge of the particle, and on the gradient of the square of the field intensity

$$F = \text{constant} \times v \times \frac{k-1}{k+2} \text{grad } E^2$$

where $F =$ force, $v =$ volume, $k =$ permittivity, $E =$ intensity.

Evaporation of polluted raindrops has produced unexpected effects where deliberately greased surfaces are used. Sufficient material has been collected to cause flashover on subsequent artificial wetting. Bird droppings, growths of moss and insect infestation have all caused flashover in special circumstances.

10.3 Purging processes

True self-cleaning by air flow, sometimes aided by suspended particles of relatively large size or mass, occurs on insulators and is promoted by designing the shape to maximise surface speeds, as shown in Section 10.2. Purging of deposits by water depends for its efficacy not only on the shape of the insulator but also on the manner in which wetting occurs.

Raindrops which are incident in high winds are able to remove most types of pollution because of their impact speed. Similarly, high-pressure sprays and jets act principally because of their kinetic energy, although swirling actions remove dirt from convolutions, provided these are not too deep[106]. Both these types of wetting are subject to the disadvantage that not all the water is bounced away from the surface after impact. Where cascades result, there is an enhanced risk of flashover because the creepage path is short-circuited (Section 14.3).

Light rain and drizzle will dissolve away the dangerous components of a deposit, but generally leave behind any inert matter. Although this is a beneficial process it does call for the provision of drip rings and the avoidance of shapes of profile which can support continuous streams of solution (Fig. 10.7). The biconvex shape, mentioned in Section 10.2, although ideal from the standpoint of low deposition, proved disastrous in an outdoor trial for this reason. Even light rain will produce short-circuiting of creepage if the catchment area is large and the shed spacing small: this is a probable cause of the poor performance of some large substation posts.

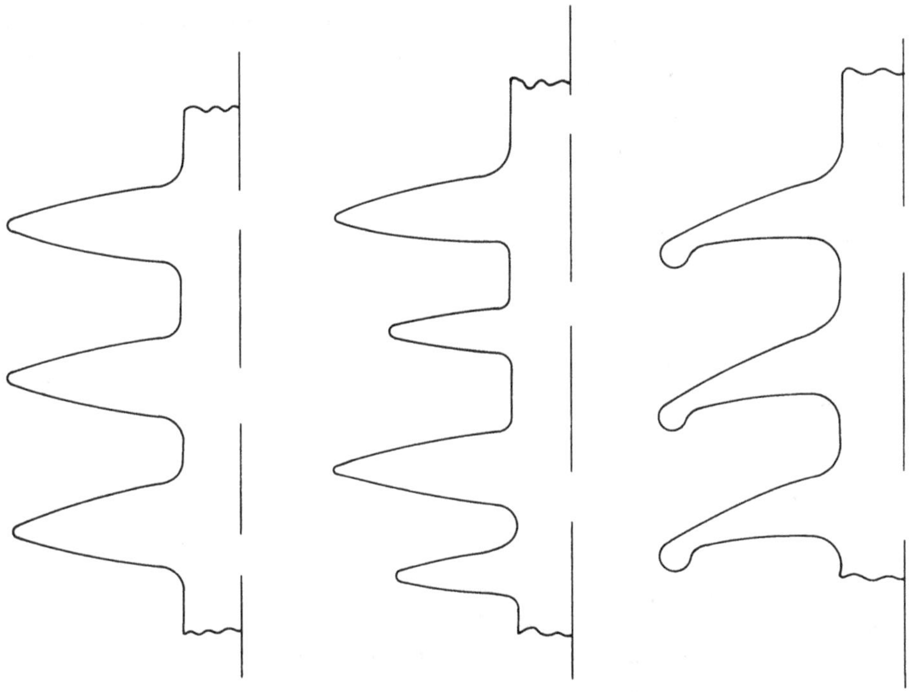

a *b* *c*

Fig. 10.7 *Conflict between aerodynamics and draining*
a Minimum pollution catch: worst draining
b Practical case: closer sheds needed to compensate for lost creepage: poor draining
c Poor aerodynamics: good draining but drips cause short-circuits in heavy wetting

Very light wetting, by dew or fog, allows solubles to be leached out of deposited layers, i.e. to migrate down the concentration gradient. Some purging does arise from this cause even in locations of negligible rainfall.

True self-cleaning is of great importance in arid regions of negligible rainfall such as the Arabian Gulf, and also in locations like railway tunnels. Insofar as purging by water is concerned there are special difficulties in countries like Japan, where there are wide variations in contamination rate, including extreme deposition in typhoons, but also advantages in a good average rainfall and the possibility of using artificial washing as a supplement[107].

A purging process which is confined to polymeric insulators is ablation of the surface. Housing materials which incorporate active fillers like alumina trihydrate do show less contamination than some others[108]: it seems that the liberation of water vapour under attack by discharges, which gives these fillers their valuable extinction function, may also dislodge deposited dirt. Other polymers which are subject to surface erosion by weathering, such as titania-loaded polyolefines, also show good ability to shed surface deposits: this is a variation on the old types of lead-based paint, which preserved a white appearance by progressively losing both surface and dirt.

Live washing as a purging process has been mentioned, but other kinds of deliberate cleaning are also used. Cleaning by hand, sometimes with acids or solvents to remove obstinate deposits, is widely practised. Initial surface treatment of the porcelain with waxes or oils aids manual cleaning, as does the use of a cold-setting silicone elastomer[109,110]. Various dry-blast processes are employed, generally on lower-voltage insulators, where relatively soft abrasives like powdered nut shells have to be employed to avoid glaze damage[111]. All kinds of deliberate cleaning are expensive in manpower and outage time. Shapes or surfaces which are effective in lengthening the period between cleanings may therefore command higher purchase prices than the normal[112].

10.4 Equilibrium deposit

On any insulator an equilibrium deposit is eventually reached when rates of catch and purging are equal, and beyond which variations are random rather than monotonic. Times to equilibrium may vary from days to years: well documented Japanese work[113] suggests a law of form

$$M = A \log t + B$$

where M = mass of deposit, t = time and A,B are constants. This work shows the strong effects of insulator shape, type, size and mounting attitude on the value of M. It finds that longrods which are easily accessible to natural washing equilibrate at low levels, but saturate in typhoons. Their deposit densities are about the same as on the top surfaces of disc insulators. Higher levels are found for the undersides of discs, while deposit densities fall as the diameter of core increases, on large cylinders. These results are consistent with Section 10.2.

Mounting attitude also produces large effects, a horizontal tension string showing much lower levels than a vertical suspension string.

More recent work by the author and his colleagues showed that quite a small inclination of an insulator from the vertical, by a few degrees in some cases, produces as much improvement in electrical performance as the full right angle[114]. This is probably an effect of lower equilibrium deposit produced by better water purging, but the air-flow will also change with slope.

The relative positions of an insulator and its main source of contamination have large effects on the equilibrium deposit. The Japanese work supports a power-law decline in severity with increasing distance from the sea:

$$M = a\,x^{-b}$$

where x is the distance and a,b are constants for different districts. Work in England shows, however, that smooth declines are far from universal and that 'skip effects' arise, whereby lower severities are found closer to the seashore than a few hundreds of metres inland. Other differences from Japanese results have been found in the consequences of pointing the undersides of horizontal strings towards the sea. The Japanese results were that heavier deposits occurred, the English work showed the opposite[115]. The simple aerodynamic results in Section 10.2 provide a plausible answer: the underside facing the flow does not favour vortex generation but promotes direct rain washing.

Common experience is that the equilibrium deposit falls off quite rapidly as the mounting height of the insulator increases. Gravitational gradients give larger densities of airborne pollutant near the ground, but the wind speeds are also lower there. Better rain washing at high level seems the probable explanation.

10.5 Assessment of required insulation: Severity measurement

The previous discussion shows that the mass of potentially conductive pollution on the surface of an insulator will generally increase with time, roughly logarithmically. Superimposed on the increase will be large fluctuations, resulting from the various purging processes and from bursts of contamination. The deposit will be challenged, intermittently, by the arrival of water, or of wet pollution, which will render conductive much or all of the soluble part of the layer (Fig. 10.8; see Chapter 14).

Leakage current will flow, when the deposit is made conductive, in pulsating form because of dry-band formation (Section 11.2), the amplitudes of the pulses rising as the surface conductance increases. At a level of conductance which is high enough to allow discharges to propagate, flashover becomes probable: this level will depend on the creepage path length.

When a new line or substation is being designed, there will be a requirement to maintain the flashover frequency below a given level. It will thus be necessary to assess the level of insulation which will suffice, for which purpose both the behaviour of the insulators and the probable rate of increase of surface deposit with time should be known. 'Severity of pollution' is the term normally used to characterise the rate of increase, although it is seen to be somewhat imprecise since one parameter cannot comprise both the equilibrium deposit and the frequency and severity of the challenges.

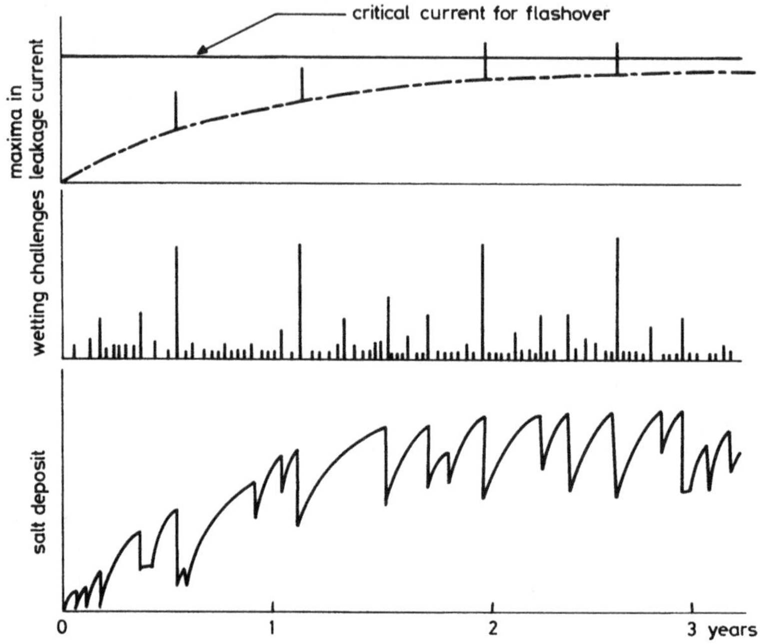

Fig. 10.8 *Schematic history of polluted insulator*
In heavy marine pollution an insulator will pass leakage current surges roughly at these annual rates:
at 25 mA: 10^6; at 150 mA: 10^3; at 300 mA: 1

Table 10.1 *Insulator zoning scheme: RWE, West Germany*

Class of insulation	Zone	Specific creepage mm/kV system
A	Agricultural, forestry; No industry	17–20
B	Light pollution, frequent heavy fog	22–25
C	Severe industrial pollution	26–32
D	Severe conductive pollution: close to large power station, chemical or metallurgical works	⩾38

Engineering practice, on assessment of required insulation and on measurement of severity, has varied widely between countries and supply authorities. At one extreme, as with the 400 kV standard construction in Great Britain, the policy, broadly, has been to insulate everywhere to a level sufficient to cope with

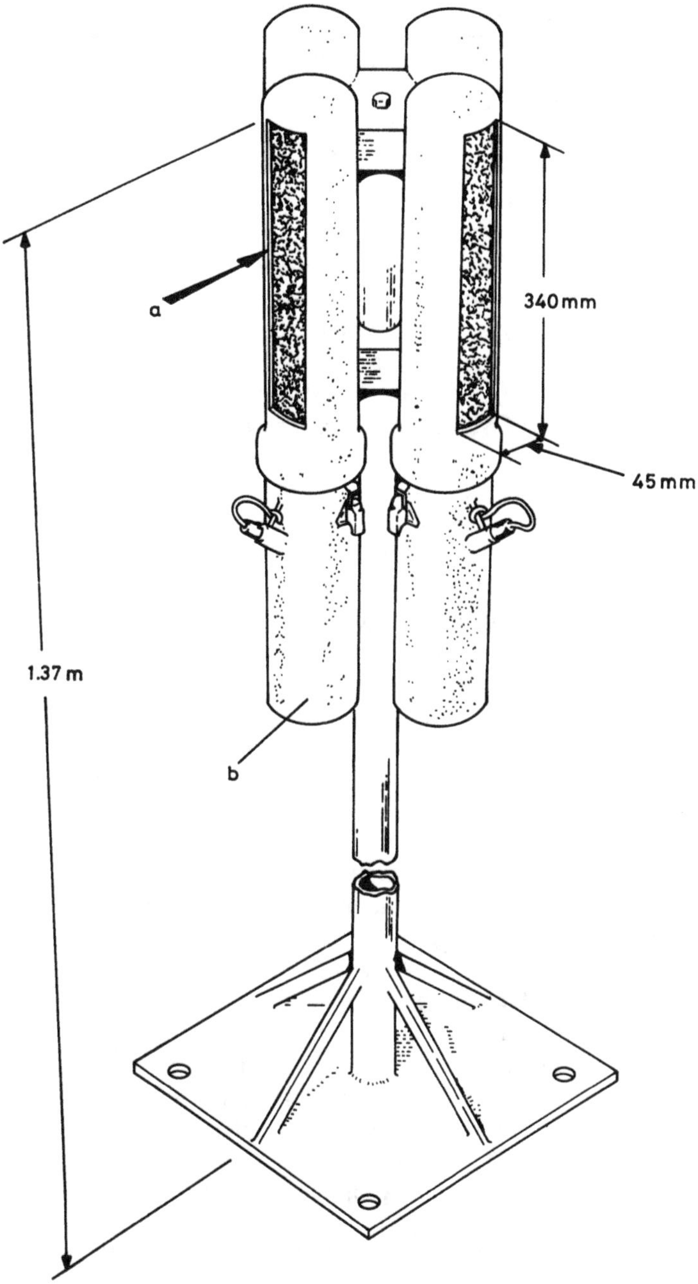

Fig. 10.9 *Directional deposit gauge (after Reference 175)*
Dirt enters at ports (*a*)
Container (*b*) can be removed and sealed for transport

Table 10.2 *Pollution severity measurement: national practices*

Country	Practices	No. of levels
Czechoslovakia	Samples used to measure, per day × cm^2, total deposit, soluble deposit; conductivity of 0·2% solution of deposit. Product of 99·5% probabilities gives level of severity	4
France	'Minipostes', having 36 kV strings of insulators, give local severity based on $I_{highest}$ ≡ max. value of leakage current	3
Denmark, Sweden, West Germany	(i) Matching locally observed $I_{highest}$ to $I_{max.}$ for a given shape of insulator (ii) Cylindrical porcelain, intermittently energised to 3 kV, giving surface conductivity (iii) Unenergised insulators as collectors, giving ESDD ≡ equiv. salt-deposit density	3
Italy	(i) Surface conductivity measured on 3 types of insulator, unenergised normally (ii) $I_{highest}$ on energised insulators	7
Japan	ESDD on standard disc insulators	3–5
Netherlands	Deduction from performance of existing installations	
UK	Generally – no differentiation of type with severity; high overall level of insulation. Some measurements with gauges, $I_{highest}$ on insulators, flashover frequency v. stress	1 + remedial action
USA, Canada	ESDD on insulators. Clean fog tests to give reference 50% flashover value	
USSR	Deduction from performance of existing insulators. Statistical analysis of maximum local surface conductivities	
Australia	ESDD, gauges, flashover frequency v. stress. Critical pollution severity compared with calculated *P*-number	4

the worst case anywhere. This has led to long strings, massive towers and high substations, but has generally been successful. At the other extreme, zones of severity are designated, with different levels for each; the numbers of zones vary from three, in France, to as many as seven in Italy. It is hard to see either technical justification for multiple zones or economic sense in the required multiplicity of designs and types of insulator and support.

Some zoning schemes are based purely on assumed correlation between type of locality and severity of pollution, such as that used by the German RWE authority (Table 10.1); others use indices of severity measured in different ways, commonly by regularly sampling deposits on real or dummy insulators or by catching airborne matter in deposit gauges (Fig. 10.9). Some electrical methods deduce the severity from leakage-current observations or by periodically immersing a target insulator in a liquid whose conductivity can be measured[115–119]. Details of many types of severity measurement are given in a survey and report by Sforzini[120]. This is summarised in Table 10.2.

At the present time the outcome of an assessment of severity is usually limited to the specification of minimum creepage lengths (for example, in mm/kV system, minima between 17 and 38: Table 10.1). However, since not all insulators use their creepage path with equal efficacy, and since it is now possible to assign numbers to different types of insulator which indicate their abilities to withstand pollution at different severities, a more logical step would be to specify withstand salinity (or its equivalent in other artificial tests) rather than specific creepage length. Details of these tests are given in later Chapters.

The basic work for matching insulators to localities of different severity was done in collaboration between the supply authorities of Great Britain, Italy and France[121,122] with the aim of using withstand salinity to characterise the insulators rather than mere specific creepage. A few new installations, for example in Sonelgaz, Algeria, are based on this philosophy: others, however, cling to creepage or specify both creepage and type, e.g. 'minimum 45 mm/kV, longrods inland, cap-and-pin, open type, on coastal lines'[123].

There is little doubt that the most used, and at present the best, method of assessing insulation requirements is to take careful note of the actual behaviour and history of existing lines in the area under investigation. Since pollution performance is accurately linear, at least up to the 750 kV level, data for new lines may be extrapolated from results of installations operating at lower voltages.

Chapter 11

Physics of pollution flashover

11.1 Flashover paradox

The apparent paradox in pollution flashover is that catastrophic electrical discharges are produced, spanning up to metres of air, by electrical potential differences capable, in ordinary circumstances, of being contained by air clearances of the order of a few centimetres. In some way, the presence of feebly conducting deposits, on a surface which otherwise would be highly insulating, lowers the effective electric strength of the surface by a factor not far short of 100.

The underlying causes are two: the localised evaporation of a film of electrolyte gives rise to breaks in the conductive film – the so-called dry bands – across which electric stresses sufficient to ionise the air are generated; arcs in a gas, once established, can readily be extended without extinction by relatively slow separation of the electrodes between which they burn.

For much of its life an insulator will run with dry bands on its surface which are intermittently spanned by discharges (Fig. 11.1). These discharges are harmless, apart from questions of possible interference generation and surface damage. Only very rarely will the combination of conductivity and electrical stress be sufficient to allow an arc to develop having sufficient current to make it self-sustaining under propagation: flashover then occurs. The technical hazard is that the surface conductivity which causes a flashover persists after the arc has been cleared by operation of the protection, allowing subsequent flashovers to occur. A reduced system security will persist until the causes are removed. It is this unique property of pollution flashover as a source of power-line outage which has attracted such interest over so many years.

11.2 Stages of the flashover process

Common precursors of pollution flashover on insulators in actual service are the following:

(a) Arrival of nearly pure water, as dew, rain or mist, at an insulator which carries a burden of pollution comprising soluble ionic components like common salt
(b) Deposition of droplets from marine or industrial fogs, or of other combinations of water and electrolyte
(c) Build-up of hoar frost, freezing fog or ice on the fouled surface of an insulator, the ionic components of the fouling then proceeding to depress the freezing point of the water and allow solution at the interface
(d) Switching in of a circuit containing insulators which are wet and fouled
(e) Arrival of a temporary overvoltage, or of a switching surge, at an insulator which is wet, fouled and possibly already energised.

Fig. 11.1 *Discharges across dry bands in natural pollution*

Of these cases, (*a*) is the most common. Especially in desert areas, pollution-flashover occurrences are closely correlated with times of dew and morning mist, while in marine-polluted regions the dangerous times are in still-air fog. Simultaneous deposit of water and solute occurs in on-shore storms and, rarely, when insulators are immersed in chimney plumes. The selection of this case, (*b*), for salt-fog testing therefore departs from generality.

The freezing-fog condition (*c*) has given rise to some of the most serious incidents; for example, in 1962 to multiple failures and the temporary break-up of the English transmission network. The offending layer of electrolyte is effectively sealed onto the insulator and requires manual removal. The remaining cases, (*d*) and (*e*), though less common, throw an interesting light on the flashover processes and are dealt with below.

We now consider the following broad stages of the flashover process: initial behaviour of electrolytic films under local electric stress; stability of discharges between different parts of the layer; propagation of discharges along the surface and evolution of an arc which short-circuits the surface.

11.2.1 Electrolytic layers under electrical stress

The electrical conductivity of an electrolyte depends on the concentration of ions, their mobilities and their charges. For a common electrolyte like NaCl solution, with two monovalent species having mobilities u, v, charge e and concentration n, the conductivity K is given by

$$K = n(u + v)e$$

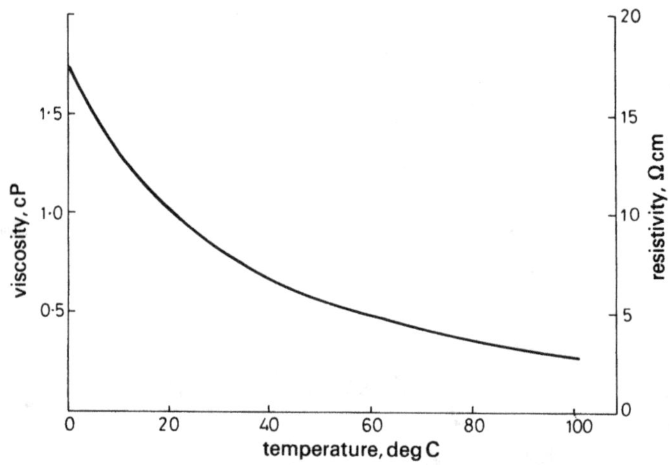

Fig. 11.2 *Electrolytic resistivity and viscosity versus temperature*
The viscosity of water and the resistivity of NaCl solution (75 kg/m^3) vary similarly with temperature

Since the viscosity of water falls rapidly with temperature increase, the mobility increases correspondingly (Fig. 11.2). At high electric-field intensities there is increased mobility arising from the stripping of water molecules from hydrated ions: this, the Wien effect, may enhance the conductivity around the ends of discharges[124].

In the case of a film of salt solution on a real insulator, the steps are as follows. Leakage current first raises the conductivity thermally. Water is lost by evaporation but with little fall in conductivity (except for the reduced dissociation with increasing concentration of solute) until solute is precipitated. Sharp local rises in local resistance and heating power then occur, with a runaway increase in drying rate. The result is a band of high surface resistivity – the dry band – which is spanned intermittently by discharges, the current in which is governed by the resistance of the remaining wet layer.

Because of the high latent heat of vaporisation of water, 2500 J/g, the bursts of pulsating current have small drying effect in comparison with the continuous sinusoidal current before dry banding. An insulator therefore spends much of its life in a quasi-stable state of surface discharge activity, with small water deposition balanced by small drying power generally, although large in the dry bands themselves. Even at this stage a mathematical model is seen to be difficult of development; experimental work is essential and was begun, mainly in England, in the 1960s.

The behaviour of layers of artificial pollution under electric stress was investigated, using strip geometry, by Hampton[125]. Strips of glass were coated with layers of kieselguhr, dextrin and salt. The voltage distribution as a function of time could be measured by capacitively coupled probes, attached to the back of the glass, one behind each terminal electrode and the remaining eight distributed along the layer. The layer was wetted by water fog, in a cabinet, and the development of flashover from dry-band discharges was monitored. The process comprised the following steps (Fig. 11.3): uniform voltage gradient; development of more than one dry band; dominance of one band; passage of arc across band; development of arcs; flashover.

Important features were the quasi-stable phase of the dry-banded surface, during which any droplets deposited within the bands were rapidly evaporated by bursts of discharge current, and the ability of the extending arc, once propagation along the surface had started, to survive extinction at current zero by restrike on the subsequent half-cycle of voltage.

The crucial unknown fact was this: what determines whether the dry band will simply persist or whether the discharges will propagate along the surface to cause flashover? Hampton showed that the criterion for propagation was equality between the stress in the arc, E_a and the stress in the layer, E_p. In general terms, as long as E_a exceeds E_p, any physical extension of the arc must lead to a reduction in current, with a further increase in E_a, since arcs show a falling characteristic of stress against current:

$$E_a = AI^{-n}$$

where I is the current and A,n are constants[126]: the extension thus ceases.

Hampton measured the stress/current relation for arcs, both in steam and in air, and investigated the propagation criterion. The arc characteristics (Fig. 11.4) show the large effect of water vapour, because of its dissociation and the large thermal conductivity of hydrogen, upon E_a. The propagation criterion was investigated directly, using a water column to represent the polluted surface

(Fig. 11.5). The oscillating alumina tube forced the arc to extend and contract periodically, the excursions leading to the introduction of an alternating component into the otherwise steady direct arc current. However, this alternating current would disappear once the voltage gradients in the arc and in the water column were equal, since the total conducting length (arc + column) was kept constant. Hampton showed that, with rising current, the alternating current disappeared at the onset of flashover, thus demonstrating the criterion $E_a = E_p \equiv$ criterion for propagation.

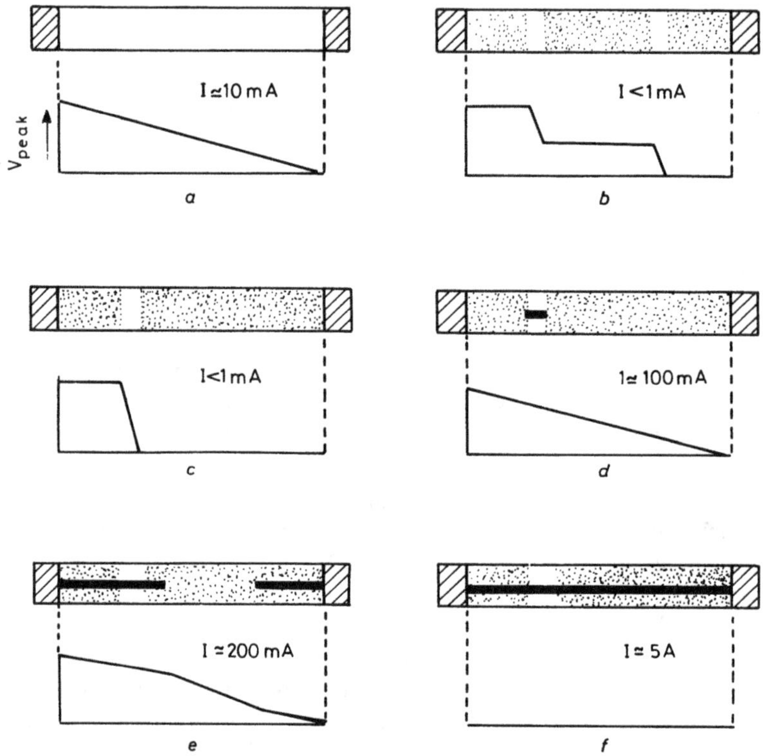

Fig. 11.3 *Development of flashover on polluted strip: voltage distributions (after Reference 175)*
a Start of wetting
b Formation of dry bands
c One band dominates
d Sparkover of band
e Extension of discharges
f Completion of flashover

This work established directly the *necessary* conditions for the start of propagation of a surface arc to flashover, a result which could be deduced from theoretical studies by Obenaus[127] and Neumaerker[128]. The condition was also consistent with the results of Alston and Zoledziowski[129]. Hampton's work, however, was never intended to comprise all the *sufficient* conditions for development of real flashover across a real insulator. This fact is obscured in the important review and analysis of flashover models by Rizk[130], summarised in Table 11.1.

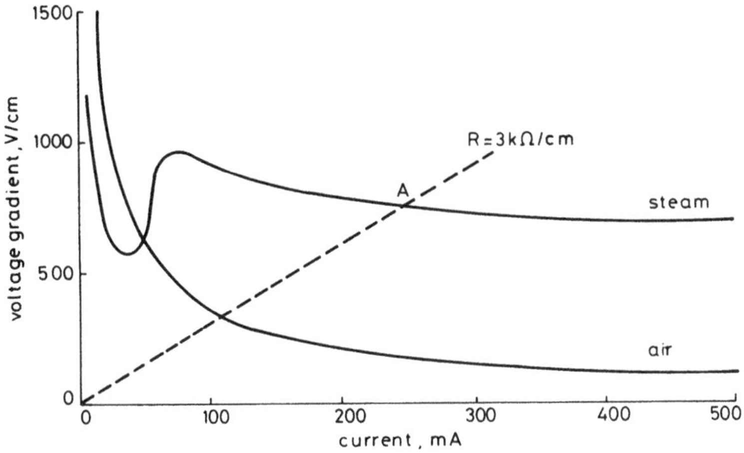

Fig. 11.4 *Voltage gradients in arc columns (after Reference 175)*
Instability sets in, for series resistor of 3 kΩ/cm, at A, for arc in steam

It is worth emphasis that the basic Hampton criterion gives much of what is needed in the design or choice of insulators for operation in pollution. It shows that flashover cannot occur if the surface electric gradient and the leakage current are sufficiently low. Their reduction certainly will prevent flashover; all other ways of interfering with propagation of discharges are, by comparison, mere palliatives.

The models and empirical theories (Section 11.3) are mainly valuable in helping to determine which parameters must be fixed to make artificial tests valid, and in particular, to set proper values of source impedance, regulation, ratio of resistance to reactance and to show the effects of resonance.

11.2.2 Propagation of discharges

A thorough investigation of the propagation process was made, using water baths, by Boylett and Maclean[131]. Subsequent work on the effects of cover plates and metallic barriers was made by Marsh[132] and by Swift[133].

The arrangement of Boylett allowed both optical and probe measurements to be made: the varying shapes of discharge with position, time and polarity were established, as well as the electrical conditions both in the discharges and in the

Table 11.1 *Stability of surface arcs: Principal results (after Rizk*[130]*)*

Author	Model	Results
Obenaus[127]	Arc of length x in series with resistance R_p of wet pollution layer. Source voltage U; current i; n and N constants.	$V_{arc} = xN/i^n$ $x = (i^n/N)(U - IR_p)$
Neumaerker[128] Rizk[130] Alston and Zoledziowski[129]	Uniform pollution resistance $\bar{r}_p$ per unit leakage path, L; arc bridges x/L; minimum sustaining voltage U_{cx}; critical stress E_c; critical current i_c	$R_p = \bar{r}_p(L - x)$ $(nN/\bar{r}_p)[U_{cx}/(1 + n)NL]^{(n+1)/n}$ $= \left(\frac{x}{L}\right)^{1/n} - \left(\frac{x}{L}\right)^{(n+1)/n}$ $E_c = N^{1/n+1}\bar{r}_p^{n/(n+1)}$: $i_c = (N/\bar{r}_p)^{1/(n+1)}$
Woodson and McElroy[138]	Disc of pollution resistance between inner and outer ring electrodes, r_i, r_o; arc spans outwards to r_a; surface conductivity K_s	$R_p = \frac{1{\cdot}6 \times 10^{-2}}{K_s}(r_o - r_a)^{1{\cdot}4}$
Hurley and Limbourn[135]	Empirical relationship between minimum AC arc voltage for rod gap x, in series with R_p; strike distance L_a	$U_c = \text{const.}\ \bar{r}_p^{1/3} L_a^{2/3} L^{1/3}$
Claverie and Porcheron[122]	Minimum arc reignition voltage for given current and distance, i and x	$U_{cx} = \frac{800x}{\sqrt{i}}$

water layers. The experiments (Figs. 11.6 and 11.7) showed the following important details:

(*a*) The simplifying assumption that the propagating arc makes a single connection to the pollution layer is incorrect: for both polarities, multiple contacts are established between discharge and conducting surface

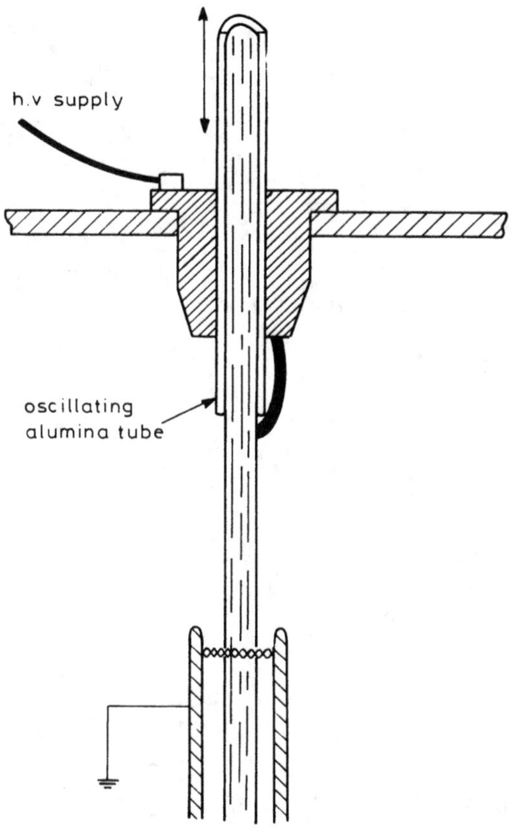

Fig. 11.5 *The oscillating arc (Reference 125)*
Hampton's experiment to determine the propagation criterion

(*b*) Both shape and velocity of propagation are strongly polarity dependent, with positive up to ten times faster than negative, for given starting conditions
(*c*) The positive discharge is constricted and has many branches: the negative is broader and simpler
(*c*) The field intensity within the discharge increases towards the advancing tip; i.e. there is no single value, of E_a, but a range of values
(*e*) The velocity alters as the resistance of the electrolyte alters, falling with increasing resistance.

Realistic values of electric stress were used in this work, i.e. between 350 and 500 V/cm, for which average speeds, from start of propagation to flashover, of between about 10 and 330 m/sec, were observed, dependent on polarity as well as starting stress. A steady direct voltage was used, supplied by a large capacitor;

Fig. 11.6 *Schlieren photograph of positive surface discharge*
The constricted positive discharge makes multiple contacts from many branches, each ending at a wavelet.
The current density evidently falls with distance from source

it is, however, clear from the relatively low propagation speeds (in comparison with sparkover velocities in air) which were found that successful development of flashover on an AC energised insulator requires survival of several voltage zeros and polarity reversals, at least for 50–60 Hz cases.

The behaviour of low-current arcs on polluted surfaces, at and after current zero, has been studied by Maikopar[134], by Porcheron and Claverie[122], by Hurley and Limbourn[135], and many others[136–140]. Rizk's review considers that a model of dielectric breakdown after current zero, i.e. a reignition under electric stress, fits the facts better than an energy-breakdown model, in which sufficient hot gas and debris from dissociated gaseous conductor is assumed to remain, after extinction, to facilitate conduction renewal with rising applied stress.

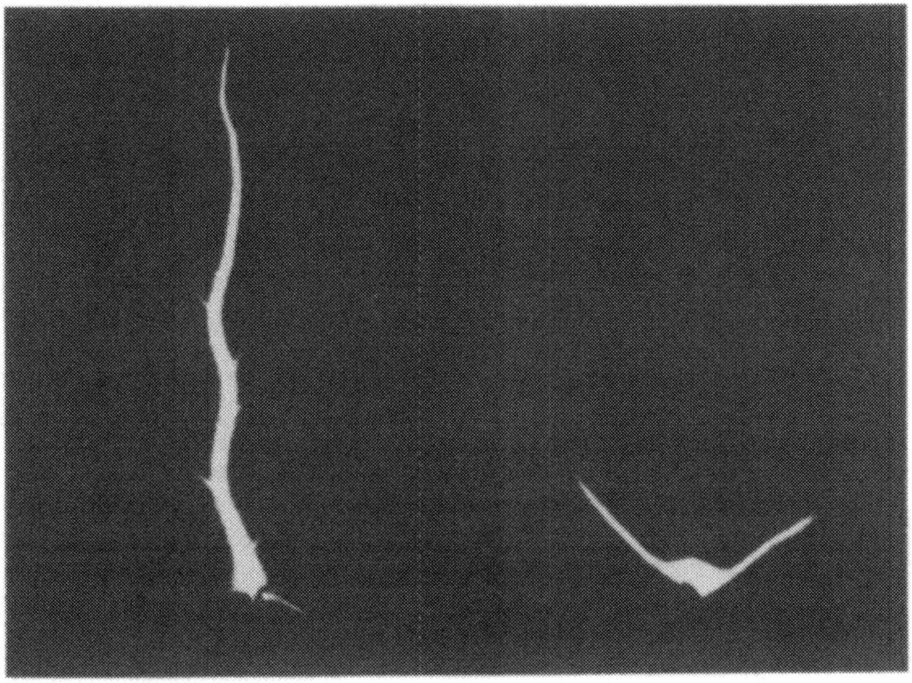

Fig. 11.7 *Discharges are strongly polarity variable*
Positive (left) discharge moves ten times faster than negative (right) and has different form.

It is sometimes found, during artificial-pollution testing in particular, that different propagation mechanisms operate in sequence during a given flashover. Visually this is often evidenced by a blue-coloured spark accompanying the normal sodium-yellow arc and audibly by a markedly different noise of flashover. Air breakdown is responsible for these effects: Swift[141] has shown that ejection of water drops into highly stressed regions of insulator surfaces may have the ability to trigger a spark which does not follow the surface, but passes between skirts or metal parts which are at large relative potentials.

As soon as any substantial length of arc has developed, even well before the end of the propagation to flashover, the path begins to be affected by extraneous influences like convection, thermal contact with solid surfaces and presence of ablated material. Where the arc passes between closely spaced surfaces[48] it may be forced into oxbow paths or even ejected by arc-chute action arising from steam or other vapours. Real improvements in flashover voltage are obtainable in this way[142].

11.2.3 Voltage waveshape and propagation

It is evident that the case where voltage is applied continuously must be most favourable for flashover, in given polluting conditions, since there is no electrical reason for propagation, once started, not to proceed to flashover. Under power-frequency alternating voltage, with current zeros, cyclic variations of stress and polarity reversals, flashover must be more difficult, since the propagation velocities are quite slow and several cycles must therefore be occupied from initiation to completion of flashover. An electrical event of duration shorter than several periods of power-frequency alternation should be capable of causing pollution flashover only if its amplitude is sufficient to confer very high speeds of propagation to the arc.

Experiments support these hypotheses. Flashover in given conditions does become more difficult as the voltage duration is reduced. In practical terms, if the power-frequency flashover voltage is V (RMS), the DC value will be somewhat lower than $\sqrt{2}V$, and the peak values for switching impulses will be higher, depending on their duration and polarity.

Pollution testing under direct voltage is expensive because of the needs for a voltage source of low impedance and ripple level, but work in England, Germany, Scandinavia and Japan[143–146] shows that values of F = (peak AC flashover voltage)/(DC flashover voltage) may exceed 1·7 for convoluted discs. The adoption of $F = 2$ gave successful results on marine polluted glass discs, while $F < 1{\cdot}2$ has led, elsewhere, to unacceptably frequent flashover.

Useful comparisons were made by Ely and Roberts[147] using switching surges. Different shapes were polluted with fogs of salinities up to 220 kg/m^3 and their 50% flashover values compared, with and without prior AC energisation, the purpose of which was to develop a system of dry bands. On anti-fog discs a waveshape of 35/2200 μs required a peak value twice that of the power-frequency voltage, with positive excursion, and some 25% more with negative. The increase was more with short tail, 2200 μs, than with longer, 3200 μs; a reduction in electrical duration of about 35% increased the flashover voltage by some 6–10%. It was shown that the presence of a family of dry bands reduced the switching-surge value by some 10%.

Results broadly in line with these were obtained in Japan[148] and elsewhere[149]. However, observations by Ely and Roberts at low severities of pollution showed major differences in behaviour with polarity, even at wavetail lengths of 2200 μs, with the positive flashover paths running clear of the insulator surfaces. It must, therefore, be the case that processes other than propagation over wet surfaces become increasingly dominant, with falling duration and increasing amplitude of electrical event. The behaviour under lightning impulses, with total durations less than 1 ms, is consistent with the assumption that electrostatic effects of water are then more important than surface propagation. Only small changes in 1/50 μs flashover value were seen when insulators, initially wet, then polluted, were compared, although very heavy pollution with associated dry bands did produce larger changes[150].

The presence of metal parts in the leakage path has important effects both on dry-band formation and propagation to flashover, not all of which are well understood. Common experience is that frontiers with metal are preferred sites for discharge activity, not only on cap-and-pin discs, where they are obviously coincident with maxima in leakage-current density, but also on longrods. Electrochemical effects operate in these places, modifying the surface conductivity; discharges will also anchor readily onto metal, where the anode and cathode falls are smaller than onto water, and where the electrode cannot retreat by drying off. Dry bands adjacent to metal are thus to be expected.

The modifications in arc propagation, investigated by Swift[133] and showing themselves in Verma's measurements, of differences in interaction between insulator and source in salt-fog tests[261], seem to depend on the thermal capacity of the metal. Recent results from work at the Brighton Insulator Testing Station appear to show that a large proportion of metal in a given creepage-path length may even outweigh the loss of insulating surface, in some cases where anti-fog discs are compared with standard profiles. This result would be consistent with the results of Kjølby[198].

11.3 Models and empirical theories of complete flashover

Following the demonstration by Forrest[151], as early as 1942, that dry bands are the precursors of flashover, models have been developed on the basis of arc stability. Resistors of different geometries have been postulated: Woodson and McElroy[138] and Claverie and Porcheron[122] examined the consequences of departure from the simplest possible strip model of resistor towards those which, like real insulator surfaces, have wide variations in conducting area and in which current flows radially rather than linearly.

Gross simplifications were generally made, especially in assumptions of single-point terminations for the arcs and uniform gradients of electric intensity within them, independent of polarity. Boylett and Maclean's[131] results show that neither is correct. Arc-quenching effects, both of water vapour and proximity of cooling surfaces, which differ markedly from shape to shape, have also been disregarded together with questions of electrostatic instability of surface-water films. Numerical constants have had to be introduced to get, in most cases, even approximate agreement with real events.

Erler[152], however, successfully explained the apparent departure from common sense in the behaviour of large-diameter insulators. Since the series resistance must fall directly with circumference increases, and thus the arc stress as well, the flashover performance should decline continuously. The flattening out which, in fact, occurs about 0·35 m diameters arises from multiple arcs across a given dry band. The periphery is effectively divided into independent strips.

Hurley and Limbourn[135] considered both arc stability and some effects of intervention by arc barriers of dielectric. Their work led to a complete philosophy of insulator shape, the *P*-number concept based on matching a number, for the shape, based on its creepage, strike distance and degree of convolution to a critical value derived from the local severity of pollution. Details are given in Section 13.1.3.2.

Another attempt to extract basic principles from studies of insulator behaviour was made, with great ingenuity, by M.P. Verma[153], in what he calls the I_{max} approach. From studies of current records up to complete flashover, Verma found that a critical peak current was reached, about 10 ms before complete flashover occurred, indicative of the imminence of flashover. This he nominated I_{max}; if it were not reached flashover could not occur, but if it were, from whatever combination of circumstances, flashover would be inevitable.

The practical applicability, and even the physical basis, of I_{max} has been criticised, but there is little doubt that useful insights into the flashover process have resulted from Verma's refined ways of measuring currents. One such, demonstrated in 1985 at FGH, Mannheim, was the use of pre-flashover current measurement as proof that the particular flashover process at work was droplet-triggered sparking and not surface propagation[154].

A further important field for empirical study has been that of bridging of convolutions in insulator profiles. Ely, Lambeth and Looms[48] established the large declines in flashover voltage which may result from drips or cascades, when these short-circuit parts of the creepage path. Previous work[75] had shown the benefits from using alternate long and short overhang in post insulators: work continues, on similar lines, today[155]. There seems little doubt, however, that attempts at compromise between the close-spaced equal-overhanging skirts, which give best results in light wetting, and the wide spaced, steeply sloped cones, which are needed for heavy wetting, are doomed to failure. A good solution is to separate the functions, using close spacing for the light-wetting case and adding water-shedding ancillaries[142] for the heavy wetting (live washing or torrential rain) case.[156]

Chapter 12

Testing of insulators

12.1 Classes of test

Tests on insulators may be classified as follows:

- For quality-control during manufacture
- To demonstrate adherence to purchasers' specifications
- For prediction of life expectancy of the insulator in service
- To determine the likely performance of the insulator when energised in pollution.

The first two categories are covered in Appendix B, and, as far as generation of nuisances like radio interference are concerned, in Chapter 16. Life expectancy is of special concern for polymeric insulators and is partly covered in Chapters 13 and 6.

An unfortunate feature of the tests which are made to meet purchasers' specifications is the quite general retention of requirements which are either useless or even counter-productive. Here may be mentioned especially power-frequency overvoltage tests, both wet and dry. These simulate conditions which are inconceivable in real service; they are incapable of detecting any structural defect and their requirements have led to the incorporation of undesirable features in some types of insulator as well as to the rejection of good, new shapes and materials.

In this Chapter attention is largely concentrated on predictive tests, aimed at determination of orders of merit, in ability to operate under contamination and at those made for purposes of research into insulator behaviour. Natural and artificial tests are covered separately.

12.2 Natural pollution testing: Background

Lundquist[3] describes in 1912 only flashover tests, respectively dry, using 2·5 to 3·0 times rated voltage and wet, between 1·5 and 2·0. Neither pollution nor any possibility of decline in performance with time was cited, omissions which largely remain after 75 years.

Designs of insulator and selection of numbers of discs must have been made on the basis of flashover experience up to the 1930s, when Forrest[158] instituted one of the first natural-pollution-testing stations. He set up insulators at a site subject to severe industrial pollution, energised them from a transformer at their rated voltage and counted the surges of leakage current at different levels. He showed large differences in cumulative surge totals between three strings of insulators, respectively standard discs, anti-fog discs and standard discs in horizontal attitude. Forrest also designed and used a probe measuring the voltage distribution along live strings, and thereby demonstrated detection of faulty units, as well as the large variations in voltage per unit which occur with humidity and contamination changes.

His technique, comprising the use of normal voltage applied to severely polluted units and deducing imminence of flashover from counting leakage-current surges, was copied both in England, at a marine polluted site near Brighton, and in continental Europe. Various expedients were adopted to reduce the measurement time: flashover frequency could be increased by raising the electric stress and surge-rate counting could be supplemented by I_{max} measurements[159–161]. An interesting observation was that quite crude telephone-type counters, using electromechanical registration of the type originally used by Forrest[158], were often better and cheaper than elaborate electronic assemblies. The latter gave more detail, but were much more vulnerable to damage and to loss of data by malfunction.

An interesting innovation by Reverey and Verma[162] was the use of automatic variation of stress at flashover. By means of fuses, insulators could be over-stressed, thus accelerating their flashover, but left in circuit afterwards with a stand-off insulator switched in by the blowing of the fuse. This scheme was developed by Lambeth, in conjunction with special explosive fuses[79], to 'float' the test voltage so as to determine the critical stress for insulators, the effective length of which was increased automatically by each fuse operation (Fig. 12.1).

The 'accelerating' methods, using either cyclic increases of voltage to cause flashover[160] or periodic applications of sustained overvoltages as high[181] as 65%, depart objectionably from reality. High-stress modes of flashover may operate and the pollution layer is frequently perturbed by flashover effects.

Since the 1960s, major advances have been made from collaboration between some European electricity supply authorities in exchanges of data from widely different test sites. In England the Brighton site eventually provided test voltages up to 1500 kV system, as well as variable stress and mechanical cycling facilities. West Thurrock, on the Thames Estuary, gave data from cement and industrial pollution. In France a severely marine polluted site at Martigues, with others elsewhere and data from miniature test points ('minipostes'), were provided. In Italy, sites were set up near Venice, in industrial and marine pollution, as well as on Sardinia and near Milan, while data from ultra-high-voltage systems was supplied from Tuscany. Close contact was also maintained with the High Voltage Institute, FGH, at Mannheim and with workers in Scandinavia and Czechoslovakia. Published work from these collaborations in CIGRE and elsewhere provides a useful and unbiased data base[35,114,121].

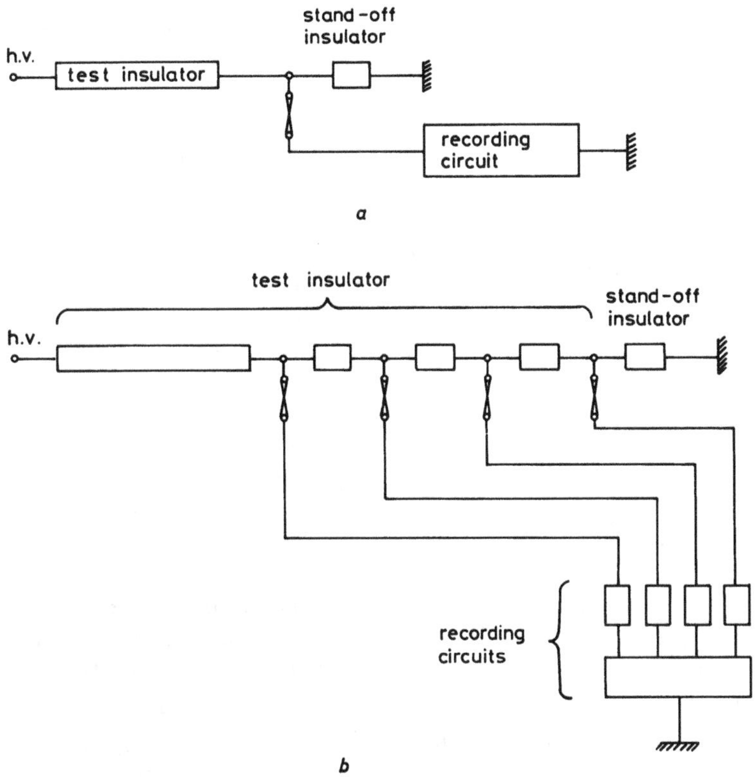

Fig. 12.1 *Fused insulator-testing circuit (after Reference 175)*
Flashover automatically increases tested length of insulator
a Single
b Multiple

In addition to the purely electrical data on flashover, outdoor test sites have provided uniquely valuable information on the deterioration of vulnerable insulating materials, especially glass, resistive glazes and polymers. Surface treatments, like greases and silicone pastes, have also been evaluated at outdoor sites. Of special value have been the tests on the secondary consequences of surface discharges[163] in causing internal damage to surge arrestors. Other work has included investigations of the ability of optical-fibre terminations, which run from live conductors to ground, to carry information in severe conditions of pollution and electrical attack[164,165] (Fig. 12.2).

12.2.1 Advantages and disadvantages of outdoor testing

As already stated, the only irrefutable system for comparing and predicting the performance of insulators would be measurement of time to flashover, at normal working electric stress and under all likely levels of pollution severity.

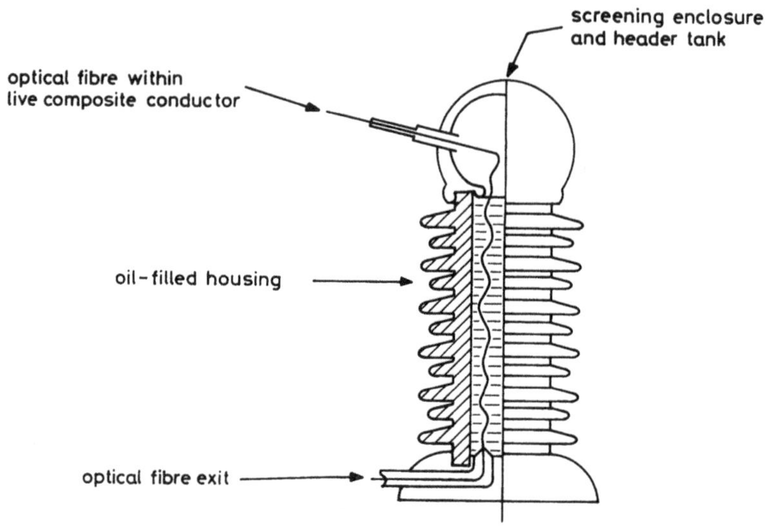

Fig. 12.2 *Oil-filled optical-fibre termination*

Since insulators are designed to work for years without flashover, such tests would be impracticable in both demands of time and number of experiments. Acceleration is a practical necessity, either by increases in severity or in stress. These, then, are the main disadvantages of outdoor testing:

(*a*) Testing in high severity of pollution sometimes gives a false order of merit. It is established, for example, that a simple-shedded post is better than a highly convoluted one at low severities but worse at high severities[166].
(*b*) Testing with enhanced electric stress both perturbs the pollution deposit by repeated flashover and raises the probability of flashover by water-drop triggering, a process which would not occur at normal stresses.
(*c*) No outdoor test, by itself, can determine whether flashover has occurred in light wetting or in torrential wetting. Entirely different profiles are needed to optimise each performance, however, and a false order of merit would be obtained if type A flashed over because of contamination whereas type B flashed over because of a fortuitous water cascade.

The substantial advantages are these:

(*d*) Outdoor testing has shown conclusively that there is no major difference between marine and industrial pollution, when shapes and creepages are to be selected. The only operative parameter is severity.

(*e*) International common programmes, under the auspices of CIGRE and other collaborations, have established orders of merit in ability to withstand pollution without flashover, for six widely used insulator types (Fig. 12.3). Artificial pollution tests have been successfully developed to reproduce this order of merit in tests needing only hours, against years for outdoor tests.

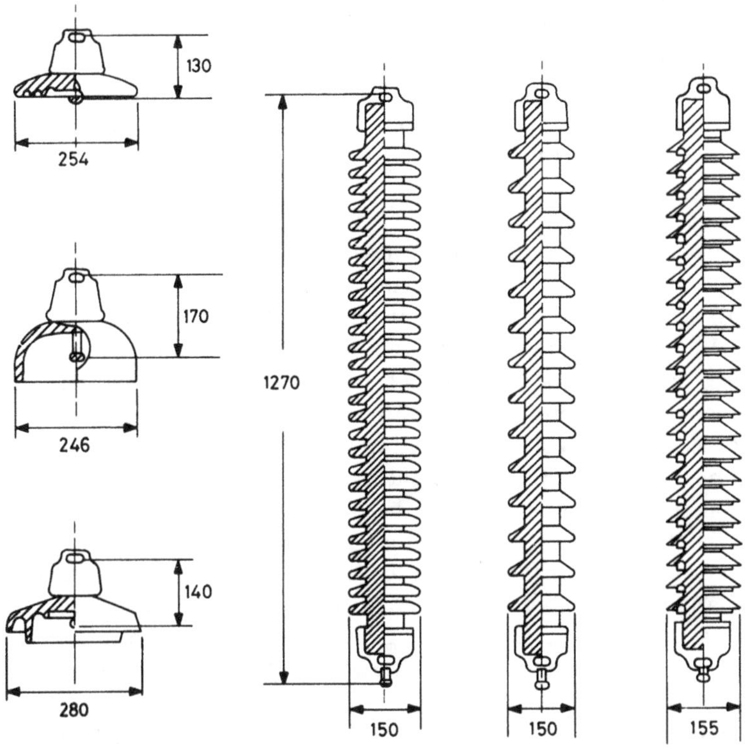

Fig. 12.3 *CIGRE insulators (after Reference 125)*
Six types, of known performance, used as bases for salt-fog test
Dimensions in mm

(*f*) Outdoor tests have shown the large variation in length of string which exists for equal flashover probabilities in severe pollution (Brighton) and negligible pollution (Leatherhead, open suburban country), in both cases with normal weathering and rain washing at work. (The ratio was no less than 9-discs/3-discs: Brighton/Leatherhead, for 132 kV system voltage).

(*g*) Outdoor tests have shown that linearity exists, in required number of discs, with system voltage, up to at least 750 kV system. This fact, that $n = kV$, is of unique value in predictions of insulation requirements for ultra-high-voltage systems where no previous service experience exists, and where errors are expensive.

When it comes to polymeric and similar insulators, having much faster surface deterioration rates than porcelain types, there is no real substitute for outdoor testing. Both the purposes and the methods are, however, different from those for porcelain. It has been found[48] that, in some cases, leakage is uncorrelated with material deterioration: major puncturing of polymeric skirts gives no detectable increase in leakage-current surging, neither does the development of cavities between moulded parts of a housing. Other materials show rapid increases in leakage and loss of flashover resistance with weathering, but are apparently little affected by discharge deterioration[35]. For such insulators, scrupulous visual inspection with, in case of doubt, X-ray internal examination, is needed at frequent intervals.

12.2.2 Practice of natural pollution testing

The basic elements of an outdoor testing station are no more than a power source of adequate capacity and instruments for observing and recording flashover and leakage. Means of gauging pollution severity are desirable.

Typical practice for a large and a small installation is shown in Fig. 12*a* and *b*, with some indications of costs; of these the dominant is that of the source, if this is a dedicated transformer. Testing power drawn from an operational line, instead of from a transformer, might seem an attractive option, but in fact the abnormal frequency of flashover, the high prospective fault currents and the lack of flexibility are usually fatal disadvantages.

Some details from real stations are illustrated: these relate to installations in the UK and are probably not typical. A climb-proof fence, warning notices and interlocks are effective in excluding unauthorised people, allowing the test insulators to be mounted below the statutory safe heights and facilitating inspection. Separate power supplies for the different voltage classes allow testing to continue, in part, when one class is shut down for work or by flashover or insulator failure.

The test insulators are mounted, at different heights as required, on gantries or towers. Either a stand-off insulator in series or a grease-protected band is provided at the earth ends, from the live side of which leakage current is conducted away to the monitoring circuitry. This counts leakage current pulses at different levels – typically 25, 150, 500 mA – and in a few cases provides continuous current traces. Elaborate electronic systems have been tried, but most were found inferior to Forrest's rugged electromechanical counters. The former provide data in saturating quantity, are vulnerable to damage and loss of data following the repeated flashovers which are designed to occur, and require expert maintenance. The latter, though crude, are adequate for purposes of comparison. Computer scanning of simultaneous electrical and environmental measurements is used in a few cases.

Accurate measurement of test voltage is essential because of the strong dependence of flashover on electric stress. Supplies are therefore either voltage-

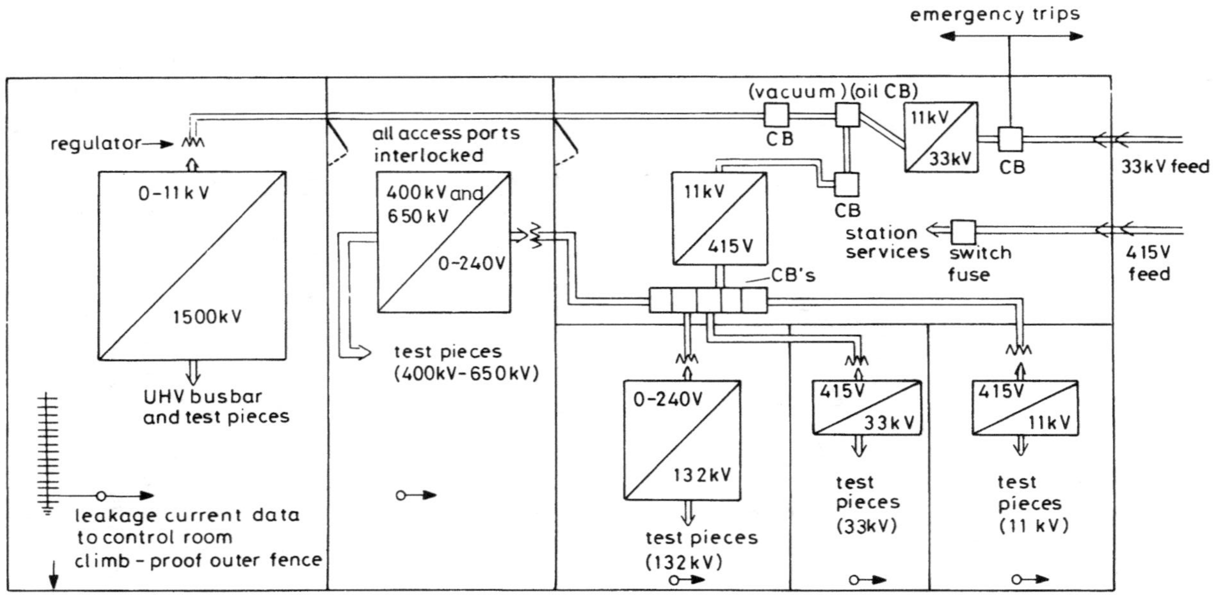

Fig. 12.4a *Circuits of large insulator testing station*
Control room: houses monitors for surge currents; meteorological data logger/correlator; voltage programmer; RI, TVI sets; alarms; switchgear
Test bays: contain pollution monitor and weather sensors; mechanical cycling rigs; hydraulic and compressed air feeds; data feeds from tested pieces to control room
Major cost items: transformers, \$M1–2; switching, protection, instrumentation, \$M1; civil works \$M0·5–1

stabilised or, where variable stress is used, subjected to careful output measurement.

Interpretation of data is not always easy, especially when different test times have to be used and where flashover does not occur on all test pieces. The different surge counts at the various levels then have to be weighted to allow extraction of comparative performances. One method is described by Lambeth[75].

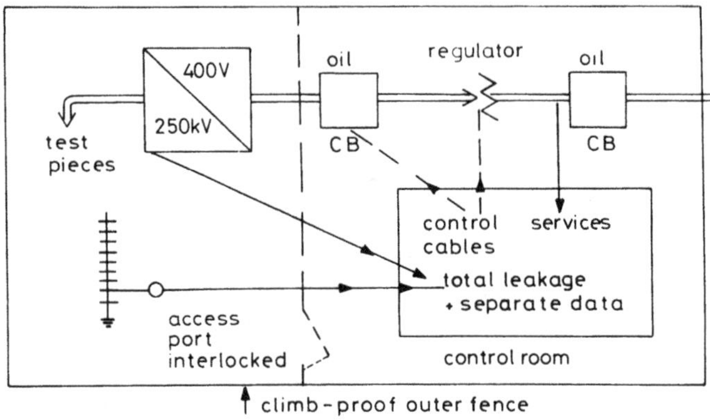

Fig. 12.4b *Circuit of small insulator testing station*
Major cost items: transformer, civil works, safety circuits if unmanned operation chosen: $M0·3–0·8

A large and costly station allows accommodation of facilities like mechanical cycling gear and interference monitors which much increase the value of the tests.

12.3 Artificial pollution testing

12.3.1 Basic philosophies

More than ten different systems for artificial pollution testing are currently in use (Table 12.1), the theoretical bases of which are not equally sound, neither are the results equally credible.

Major controversies have raged about the relative merits of some of these tests, especially where the salt-fog test is compared with those using pre-applied layers. In what follows, the author's personal opinions are expressed which sometimes differ from the claims of the actual inventors of the tests and the opinions of expert commentators. Allowance should therefore be made for the fact that the salt-fog test was developed, mainly by Ely and Lambeth[7], at the Central Electricity Research Laboratories as part of a research programme managed by the author. Prejudice in favour of that test may thus be assumed, but also an intimate knowledge both of its background and of its evolution to the status of a principal international standard.

The true purpose of the salt-fog test, underlying its invention in 1963, was to establish rapidly the relative performances of insulators having different shapes and mounting attitudes in given severities of pollution. Subsidiary aims were: to measure the functional relationships between flashover voltage and severity of pollution for different insulators; to establish laws, such as the extent of linearity of flashover voltage with length, for insulators generally; to predict the performances of new types and shapes of insulator.

The starting point for the test was the established order of merit for the six CIGRE line insulators (Section 12.2.1.). The inventors decided to adopt the normal maximum service voltage as the test voltage, with appropriately low source impedance, and went on to find a system of bombarding the insulator with atomised salt solution which would reproduce the 'true' order of merit. The variables were the number and shape of nozzle, the pressures and flow rates of atomising air and salt solution, respectively, and the range of bombardment. The adopted criterion was salinity of atomised solution which could be withstood for 60 min without flashover.

Salt solution was adopted because, as a pollutant, its conductivity could be both easily varied and accurately measured. Although the modal size of salty droplet which resulted, about 4 μm, is about the same as in natural fogs, it was never suggested that the test should 'reproduce' marine pollution or would intrinsically be more applicable to coastal than to inland and industrially polluted insulators.

It is reiterated that the choice of polluting parameters was made to reproduce a given order of merit. Apparently minor changes in flow rates, alignment of nozzles, even in surface tension of pollutant from added wetting agent, gave orders of merit which departed from the established one: these were identified and rejected during the course of international standardisation of the test.

The salt-fog test is thus seen to produce distributions of surface electrical conductivity on widely different line insulators, the behaviours of which correspond to reality in all kinds of polluting environment. The distribution is the resultant of aerodynamic catch, hydrodynamic draining, electrostatic ejection of drips and evaporation: one cannot therefore safely assume that valid predictions will be obtained for shapes differing substantially from those of the CIGRE set, or for surfaces other than glass or ordinary ceramic glaze.

Turning now from the salt-fog test to other tests, one finds that, whereas the former is dynamic, i.e. produces its deposit by continuous addition and depletion, the latter are static; they use preapplied deposits, comprising a conductive element, usually common salt, with binders and bulk formers. Electrical conductivity results from water which is either present from the start or is added, by bombardment or condensation, during the test, and which puts part or all of the NaCl into solution as an ionic conductor.

All these tests use a deposited layer of thickness as uniform as can be arranged; immediately one sees that this represents a major departure from reality in that not only are real deposits nonuniform but also that 'good'

Table 12.1 *Some artificial pollution tests*

Test	Criterion	Pollutant	Wetting	Voltage	Duration (mins)	Standard deviation (%)	Remarks
1 Salt-fog (withstand salinity)	Withstand (W/S) 3 out of 4 tests	NaCl solution	Solution directly sprayed onto test piece	Constant: at or near working level	60	2–3	Dynamic test
2 Salt-fog (flash-over voltage)	$V_{50} \equiv$ 50% flashover voltage (FOV) in given salinity	NaCl solution, 80 kg/m^3	Solution directly sprayed onto test piece	90% of FOV before wetting: raised 2% in 5 min	about 45	2	Equilibrium layer in fog
3 Kieselguhr	$V_{10} \equiv$ W/S $\equiv$ 10% FOV	Kieselguhr, dextrin/aerosil, NaCl	Condense on cooled piece: Direct spray: Steam fog:	Constant at max. conductivity	15	5–13	Initially uniform layer
4 Kaolin (steam fog)	V_{50}	Kaolin/Fullers Earth, NaCl	Gradual steam fog	Constant before wetting	about 150	}	Water added during test (rows 4–5)
5 Cement	V_{50}	Cement	Sprays, top and bottom at 10 mm/h	Constant before wetting	15–25	4–10 }	Water added during test (rows 4–5)
6 Kaolin (fog withstand)	max. W/S	Kaolin, NaCl	Indirect spray	Constant before wetting	30		

7 Cement	V_{50}	Cement	Fog spray before test	Constant after wetting	2–5		
8 Kaolin (equivalent fog)	V_5: 5% FOV from > 10 tests	Kaolin, NaCl	Pollutant sprayed on wet. Left for 3 min	Raised to FOV at 10 kV/s		3·7–7·0	Initially uniform layer. No water added during test
9 Silica flow on	FOV or W/S: V_{50} or V_{10}	Fine silica, wetting agent, NaCl	Pollutant flowed-on wet. Left for 5 min	Raised to FOV at 7.5 kV/s or to W/S value		7–10	
10 Methyl	V_{50} at one conductivity	Methyl cellulose, chalk, NaCl: aged	Pollutant sprayed on wet. Left for 30 min	Constant	5	0·3	
11 Dust spray (DC test)	FOV	Initial kaolin dust deposited (onto energised insulator) Then low-salinity fog (onto dead insulator). Then low-salinity dense fog (onto ramp-energised insulator)	First stage: none, dust only. Second stage: 0·35 l/m fog Third stage: 5·0 l/m fog	First stage: 50 kV/m of creepage Second stage: none Third stage: raised from 40 kV to FOV every 150 s	120		

W/S: withstand salinity
FOV: flashover voltage

insulators, aerodynamically speaking, collect much less deposit than bad ones. Preapplied deposit must therefore provide intrinsic, unfavourable bias when used to evaluate aerofoil, low-catch shapes. It must also fail to indicate the advantages of surfaces which are 'non-stick', hydrophobic, very smooth or even sloped, to promote draining, since all will be blanketed with the same layer in the test.

Another defect which arises from the static deposit is change of concentration with time of the NaCl. Migration occurs, at different rates in the various tests, whereby the conductive component leaches out of the layer and the known initial conductivity changes. For some of the preapplied tests, therefore, there is only a limited period of validity, perhaps a few tens of minutes, beyond which the state of the layer becomes indeterminate.

One aspect of artificial testing which is usually decisive has nothing to do with physics but with the noxious nature of salt fog itself, which is both invasive, corrosive and highly unwelcome either within testing laboratories or when exhausted to the environment. All artificial tests need powerful voltage sources which are expensive, but salt-fog containment is far more costly than the enclosure needed for steam or town-water fogs.

One advantage of the salt-fog test over all others must be mentioned; its recently discovered ability to act as a source of controlled contamination in tests also involving wetting[142]. Simulation and measurement of the effects of heavy wetting, a known source of flashover where insulators are either hot-washed or subject to torrential rain wetting, are evidently impossible with any static pollution system: they have been successfully investigated with a slightly modified salt-fog test, however, in a research programme which has disclosed the important fact that, in general, there are two orders of merit for insulators, respectively for light and heavy wetting[157].

To summarise the basic philosophies of artificial pollution testing, therefore, one can identify two categories. The proponents of the salt-fog test aim to match electrical differences in behaviour which are established as real, between different types of insulator in service and under outdoor test, by semi-empirical arrangements of salt-fog generator. The preapplied-layer proponents, on the other hand, start from various assumptions about surface deposits, in particular their probable equivalent densities of soluble electrolyte, and go on to apply such deposits and to wet them in different ways. Salt-fog deposits are dynamic and nonuniform; preapplied deposits are static and uniform initially but change their physical states, especially electrical conductivity, continuously and irreversibly during the progress of the test.

12.3.2 Principal artificial tests

12.3.2.1 Salt-fog withstand salinity test: Details are given in IEC Publication 507. Two columns of spray nozzles are used, opposed at 180° and facing the axis of the insulator from a range of 3 m. Each nozzle (Fig. 12.5) is fed with salt

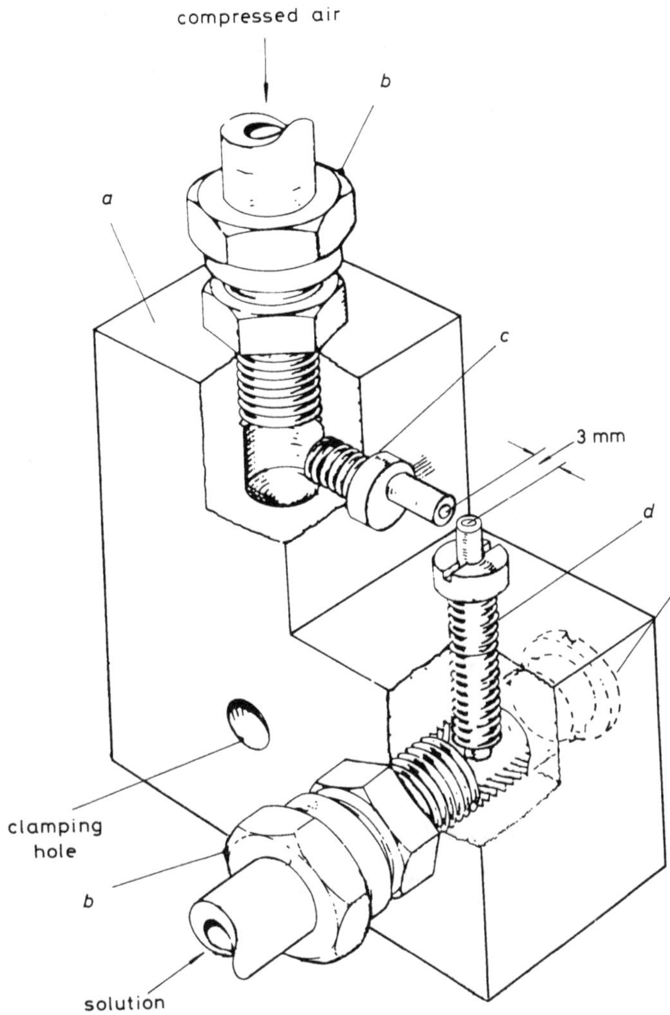

Fig. 12.5 *Atomising nozzle for salt-fog test (after Reference 175)*

a Plastics body

b Noncorroding coupling for 8 mm-bore tube

c 6 mm nominal SI thread × 16 mm-long nylon cheesehead screw with slot skimmed off and stainless-steel tube, 1·2 mm internal diameter, 3·0 mm outside diameter, fitted eccentrically to permit tube alignment

d 6 mm nominal SI thread × 16 mm-long nylon cheesehead screw with stainless-steel tube, 2·0 mm internal diameter, 3·0 mm outside diameter, fitted centrally

e Plastics plug

solution at 0·5 litre/min and with oil-free compressed air at 7 kg/cm^2. The withstand salinity of the tested insulator is defined as the highest concentration of salt, in kg/m^3, which is withstood without flashover for three out of four 1-hour tests at rated working voltage.

Additional information is sometimes obtained on the variation of flashover voltage with salinity by means of an 'up and down' series, starting from the estimated V_{50} (50% value); (Fig. 12.6).

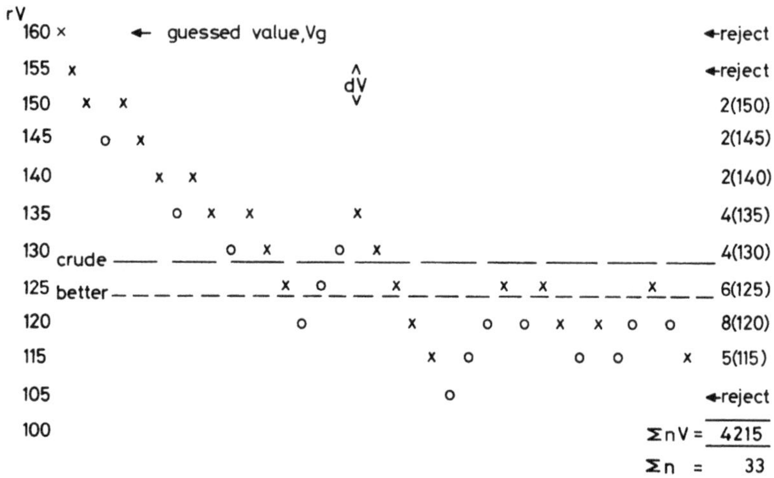

Fig. 12.6 *'Up and down' method*

× Flashover
o Withstand

1 Guess flashover voltage V_g, and select interval dV, say 3%
2 Test: if flashover, drop by dV; if withstand, raise by dV
3 Complete series; between 20 and 40 tests preferably, but 8 to 10 commonly
4 Calculate weighted average $(\Sigma nV/\Sigma n) = V_{50}$, rejecting one-shot levels

$$V_{50} = \frac{\Sigma nV}{\Sigma n} = [4215 \div 33] = 127{\cdot}7\,\text{kV}$$

Example shows clear conditioning effect; better value might follow rejection of initial two-shot levels. Crude value = 127·7 kV; better value = 123·9 kV.

For the salt-fog and all other tests a power source is required which is capable of delivering the large pre-flashover currents without serious voltage drop, or harmonic distortion in waveform. Questions of source impedance, resistive and reactive components and capacitive discharge, all of which have a bearing on the validity of artificial tests, have recently been discussed in detail by Rizk[167]: generally, however, a short-circuit current exceeding 10 A is needed, which calls for expensive transformers, voltage regulators and primary supplies, at least for tests at 400 kV and above (Table 12.3).

12.3.2.2 Pre-applied Kieselguhr test: This test was developed in Germany[168,180] and is described[181] in VDE 0448. Kieselguhr is a mineral comprising fossilised diatoms and having large surface-absorption characteristics. It is blended in water with aerosil, a silica of very small particle size, and common salt, giving a suspension which can be sprayed onto the insulator and dried.

Controlled wetting, either by atomised water or steam, is applied, with a low voltage across the insulator to allow its resistance to be monitored. When the resistance has fallen to a stable minimum the test voltage is applied. Four withstands, each of 15 min duration of test, are needed to establish the withstood surface conductivity σ. This figure is derived from the geometry of the piece, assuming uniform deposit and wetting, using the resistance value as measured at low voltage:

$$\sigma = \frac{1}{R}\text{(form factor)} = \frac{1}{R}\left(\frac{1}{\pi}\int_0^L \frac{dl}{d}\right)$$

where R = resistance, L = leakage path, d = diameter for leakage l.

12.3.2.3 Methylcellulose test: The methylcellulose test was invented in Germany, but was much developed by Dana Kohoutova, a Czech engineer, because of the prevalence, in parts of her country, of highly adhesive dusts arising from lignite-fuelled generating plant[182,183]. It illustrates the ultimate departure from reality, in pre-applied testing, by using a coating chosen for its convenience in testing but quite impossible to find, in any imaginable circumstances, on an insulator in the field.

The coating is made up by suspending in salt water an inert filler, floated chalk, with a component for controlling viscosity, methylcellulose. This material, used as a crack filler in industry, has the property of swelling and dispersing when left in water for long periods: it can then be applied to give thick and uniform coatings which lose water only slowly.

The mixture is sprayed onto the insulator, or the insulator is dip coated, after the mix has matured for two days. The coated insulator is held for not less than 30 min in a humidity chamber, after which its resistance is measured at low voltage. The test voltage is then applied for up to 5 min without additional humidification, giving either withstand or flashover. Layers are not re-used.

An evident defect of this test is the impracticability of monitoring the thickness of the pre-applied layer. One or two thinner patches will much increase the end-to-end resistance and thus give a false picture of the mean surface conductivity.

12.3.2.4 Japanese equivalent-fog method: This method, claimed to have been in use[184] since 1958, uses a layer of kaolin with NaCl, which is sprayed onto a carefully degreased and washed insulator. The layer is left to stabilise for about three minutes, after which a test voltage is applied, rising uniformly until flashover. This step is rapid: the authors speak of a 30 s rise time.

Table 12.2 *Dependence of flashover voltage on surface conductivity*

Type of pollution test	Type of insulator or model	Reference	p	Increase in flashover stress for ½ conductivity
				%
	water jet	125	0·15–0·24	10–18
	polluted plate	129	0·43	35
	water trough	169	0·2–0·25	15–19
	polluted plate	170	0·2–0·25	15–19
Kaolin, NaCl deposit and spray	cap-and-pin	171	H 0·18	13
			L 0·40	32
	rods		0·29	22
Methyl cellulose	helical-shedded rod	172	0·20	15
	rods and cap-and-pin	173	0·34–0·40	26–32
	rod: VKL 75/14	173	0·31	24
		162	0·46	37
		162	0·38	30

Kieselguhr	rod: VKL 75/14	162	0·35, 0·40, 0·60	27, 32, 51
			0·25	19
	380 kV posts	174	0·30, 0·39	23, 31
Kaolin [Fuller's Earth, NaCl]	cap-and-pin	166	0·12–0·32	9–25
Kaolin, NaCl	cap-and-pin	113	0·24	18
Kaolin, NaCl, constant voltage	bushings	165	0·16–0·32	12–25
Kaolin, NaCl, rising voltage		165	0·08–0·26	6–10
Salt fog	400 kV posts	75	0·10	7
	plain-shed posts	175	0·27–0·38	20–30
	cap-and-pin	175	0·19	14
Cyclic direct voltage	post	145	H 0·31	24
			L 0·41	33

Critical stress for flashover E_c and surface conductivity σ_s are related by $E_c = K\sigma_s^{-p}$. The index p is tabulated for some artificial tests and models

H relates to high conductivities, L to low (after Lambeth[175])

After flashover the insulator is washed and recoated for subsequent repeat tests. The equivalent salt-deposit density is determined at some time during the test, not made clear by the authors, who admit that the test results will be influenced by the place of test (indoor or out), the waiting time between coating and testing, the rate of rise of voltage, the source impedance and the kind of insulator. Even the type of kaolin, which governs the rate of loss of water, is required to be kept constant during a given test series!

Table 12.3 *Equivalent severities of natural pollution: Withstand values from different tests*

Test type →	Salt fog (kg/m^3)	Steam fog Equiv. fog Fog W/S (mg/cm^2)	Kieselguhr Flow-on Methylcell. (μS)
Natural level			
None/slight	5	0·0125	1·25
Very light	10	0·025	2·5
Light	20	0·05	5·0
Average	40	0·1	10·0
Heavy	80	0·2	20·0
Severe	160	0·4	40·0
Exceptional	224	0·8	80·0

12.3.2.5 Features of other pre-applied pollution tests: The most interesting features of other test methods are the ways in which water is applied and the relevance to real conditions.

A common and dangerous cause of flashover, especially in desert regions, is condensation of dew onto heavily salt-contaminated insulators also carrying adherent dusts. The cause of the condensation here is radiation of heat from the insulator to the night sky, a process which must depend greatly on the emissivity and thermal conductivity of the insulator and on the degree of optical shielding which its design provides. No test exists, at the time of writing, which even approximates to this condition. The Japanese fog-withstand method, of Tagaki *et al.*[185] uses a layer of kaolin and NaCl with indirect fog as the wetting process. The steam-fog method, of Kawai and Milone[186], is based on very slow deposition of water from steam. Neither takes proper account of insulator shape, the steam-fog system being also subject to criticism for the large variation of deposition rate along the tested insulator.

'Cold switch-on' flashover is a real problem which results from the need to switch out many transmission lines at times of low demand. Reclosure onto insulators, which may be wet, polluted and cold after some days of unenergised rest, commonly causes persistent flashover. This case is easily accommodated by the salt-fog test, but only the methylcellulose pre-applied test and the 'flow on' method of Macchiaroli and Turner[187–188], appear as possible alternatives.

12.4 Comparison of artificial-pollution tests

An artificial-pollution test should be valid, i.e. match service experience and natural testing, be repeatable and be reproducible in different laboratories. Comparison of some of the cited tests has been made on these three criteria in a valuable CIGRE collaboration directed by Sforzini and Schneider[120]. To date, only two of the tests – the salt-fog and the Kieselguhr – have been internationally standardised.

Some idea of the relative merits of the tests can be had from Tables 12.1 and 12.2, but the essential disparities are clear from Fig. 12.7. There is not even approximate agreement between the performances at nominally comparable severities, observing strings of anti-fog discs or longrods, in both cases, of comparable shape. It is, however, encouraging to see that almost all tests show consistent differences between the strings and longrods, and that the functional relationship (logarithms of voltage and severity plotted) is fairly linear and has about the same slope for all.

It is not easy to unify results, i.e. to compare performances as measured by different methods, because of the effects shown in Fig. 12.7. However, some success can be had by using as a yardstick the 'equivalent severities' of differently polluted places. Knowing that we need an insulator having a salt-fog withstand salinity of, say, 80 kg/m^3 to operate successfully in a heavily polluted site at 85 kV and that a similar class of insulator, at a similarly polluted site, needs to withstand 0·2 mg/cm^2, as measured in the Japanese equivalent fog test, at the same voltage, the methods can be related. Practical numbers are given in Table 12.3.

12.4.1 Severity parameter and voltage/severity function

Consideration of the simplest possible model of a polluted insulator, comprising a uniformly conductive strip of length L, between electrodes having a potential difference V, we see that the stress in the pollution E_p is independent of the surface conductivity σ and given by V/L. Assuming a small arc on the surface having internal stress E_a which is related to the current I by a power law

$$E_a = KI^{-p}$$

and applying the Hampton criterion, that arc propagation can begin when $E_a = E_p$, it is clear that, with rising surface conductivity caused by accretion of electrolyte or water, the arc current will also rise, since

$$I = V/(\text{surface resistance})$$

and E_a will correspondingly fall until it equals E_p.

The relationship between flashover voltage and surface conductivity will therefore follow the same power law as the voltage/current characteristic of the arc, provided that propagation inevitably leads to flashover which, on this simple model, it does.

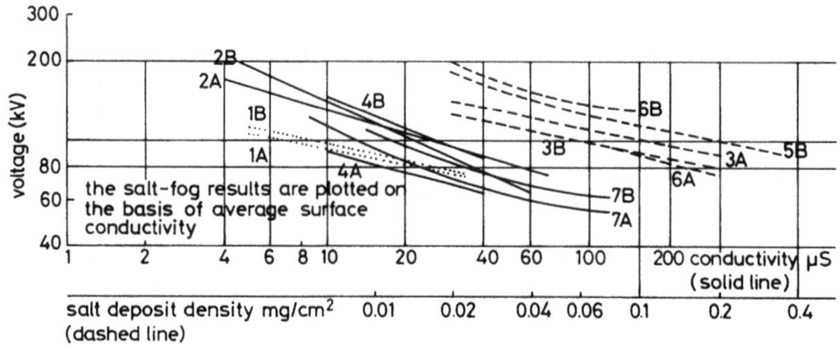

Fig. 12.7 *Comparison of artificial test procedures for nominally corresponding severities*

Salt-fog test: withstand voltage	Longrod NVKL75/27	1A
	String of discs	1B*
Methyl cellulose: mean flashover	Longrod NVKL75/27	2A
	String of discs	2B*
Equivalent fog: mean flashover	Longrods, 1·5 Jap.	3A
	String of discs	3B⊕
Kieselguhr: mean flashover	Longrod NVKL75/27	4A
	String of discs	4B*
Fog withstand: withstand voltage		
	String of discs	5B⊕
Steam fog: withstand voltage	Longrod NVKL75/27	6A
	String of discs	6B*
Wet contaminant: withstand	String of discs	7A°
	String of discs	7B†

* 9 Bullers T9336 anti-fog disc
⊕ 9 Japanese deep-rib anti-fog discs
° 9 Standard discs
† 8 Anti-fog discs
(*after Reference 120*)

The primary severity parameter is therefore the surface conductivity: we may write

$$\text{(flashover stress)} = E_p = E_a \ K'\sigma^{-p}$$

where K' is a second constant, and hence

$$\log \text{(flashover stress)} = \log K' - p \log \sigma$$

If now the surface deposit density of salt is δ and all the salt is dissolved, the surface conductivity is nearly independent of the thickness of the solution layer and directly proportional to δ.

A secondary severity parameter is thus the surface deposit density δ, and again, with a third constant K''

$$\log \text{(flashover stress)} = \log K'' - p \log \delta.$$

The relationship between conductivity per square σ and salt deposit density δ is

$$\sigma\ (\mu\text{S}) = 1{\cdot}5 \times 10^3 \delta\ (\text{mg/cm}^2 \text{ of NaCl}).$$

Direct comparison of flashover-voltage/severity-functions (Fig. 12.7) shows that simple equivalence between conductivity and deposit density does not hold. The reasons include errors of averaging, since the real deposits are never exactly uniform and often highly nonuniform in density, and incomplete wetting of deposit, leaving undissolved and thus nonconducting salt.

The slopes of the different log V/log δ functions in Fig. 12.7 show that the effective power p is not always exactly the same. Lambeth[175] has calculated values of p from a wide range of published results (Table 12.2), but it is interesting to note that Hampton's value of p, obtained for steam, applies only to part of the relevant current range and differs markedly from the value for air. Most observed values of p lie between his two limits.

Among the physical properties of real, polluted insulators which may contribute to the large observed differences between flashover/severity curves, obtained from different insulators in the same test or from the same insulator in different tests, nonuniformity in surface conductivity is likely to be dominant. The value of p for surface arcs depends on layer thickness, for uniform conductivity; flashover voltages are similarly dependent[120]. Where the conductivity is nonuniform as well as the thickness, faster propagation occurs across the highly resistive sections[131], since the surface stress is higher there, thereby raising the probability of flashover at a given overall average surface stress, assuming arc extinctions at voltage zeros. On the other hand, small regions of very high conductivity and heat absorption are known[133] to act as flashover barriers, suggesting that similar effects may arise on longrods when these carry very nonuniform deposits.

Experimental investigations of the effects of nonuniformity have, in fact, led to contradictory results. Rods with large differences between half-length coatings, in conductivity, or with alternating sections of high and low conductivity, were found to have lower flashover stresses than the uniformly coated equivalents[189]. Japanese and Russian work with disc insulators supported this effect of reduction with increase of nonuniformity[113,171]. The opposite was found by Nasser[177], who observed lower flashover voltages for uniform conductivity.

Other likely contributors to the observed variations are differences in arc path, both with changes in applied voltage and with test method, and operation of other flashover mechanisms than classical propagation of surface arcs. Some possibilities are shown in Fig. 12.8.

12.5 Source impedance: Effects on test results

The process of pollution flashover involves transitions from sparks to arcs as well as recoveries of conduction following current zeros. Both these effects depend on the rate at which energy can be supplied to the spark gap or arc path. The author's research group showed, for the case of a gap fed through cables from a large capacitor, that a local low-energy discharge precedes the main arc, the current flow in the first depending on the length and characteristic impedance of the cables and in the second on the lumped inductance and capacitance of the whole circuit[178].

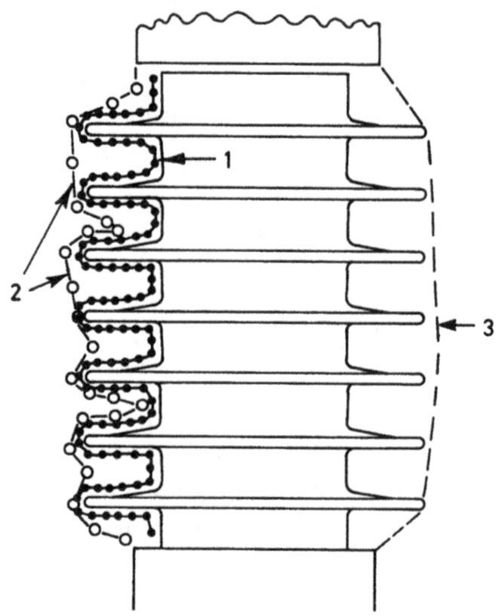

Fig. 12.8 *Different flashover mechanisms*
●–●–● 1 Classical surface-leakage flashover under light wetting
○–○–○ 2 Partial or complete bridging by drips or cascades under heavy wetting
– – – – – 3 Air-flashover by ejection of water droplets: the usual process in wet tests at elevated voltage

In real flashovers on lines or in substations, there will be some variation in behaviour on flashover because of these ignition and reignition effects, but the large currents which flow during the last stages of propagation, and which decide whether or not flashover will occur, will not be affected significantly by the source impedance.

Tests which rely on transformers to supply their power, both for outdoor, naturally polluted stations and indoor, artificial-pollution chambers, are, however, limited by economics in the regulation percentage and short-circuit current capability of the source. It is then necessary to try to define minimum-capability, i.e. cheapest, sources which will give valid test results, and to investigate the interaction of tested insulators with a variety of power-supply parameters.

Some parameters are well recognised and are already covered by international standards. These include the short-circuit current I_{sc} and the equivalent ratio of reactance to resistance X/R, both of which will evidently govern the applied voltage during the approach to flashover, when relatively large leakage currents are being drawn from the source. More complex phenomena have, however, also been identified, in particular the production of 'ringing' in the voltage waveform by excitation of resonance between the stray inductance and capacitance, and the intrusion of harmonics. The latter has had serious effects in laboratories where the test source shares a low-voltage supply circuit with other apparatus, like test lines, which reflects large capacitances into the primary supply[179].

Table 12.4 *Minimum short-circuit current ratings*

R/X		$I_{sc\ min}$
Salt-fog method	Solid-layer method	A(RMS)
0·05	0·05	5
0·05–0·15	0·05–0·15	7
0·15–0·50	0·15–0·30	10
0·50–1·50	0·30–0·50	15

After IEC Publication 507, Table 1.

Acceptable characteristics are laid down in IEC 507, including I_{sc} for different values of the ratio R/X (Table 12.4). International work on source characteristics continues, but a useful experimental and theoretical study by Rizk and his colleagues at IREQ, Canada,[167] has thrown light on a number of effects. First, on the effect of capacitance, it is established that this largely governs the shape of the leakage-current pulse. A weak source, in the sense of one having high inductive voltage drop, can be partly compensated by added shunt capacitance, although the effect differs between posts of relatively large diameter and strings of discs having 'arc stopping' intermediate metalwork. Secondly, on the effect of voltage drop caused by leakage current, 'ringing' in the voltage waveform may lead to different peak values of voltage during the critical half-cycle when propagation starts, and at a phase angle of $\pi/2$ after voltage zero; it also retards the decline in voltage during the leakage current pulse, compared with the open-circuit value (Fig. 12.9). The transient recovery rate of voltage, after current zero, is now known to be as fundamental for continuation of propagation to flashover as it is, in other fields, to the successful operation of circuit breakers.

12.6 Principles of mechanical testing

For classical overhead-line insulators, failure is readily defined. Longrods have an average breaking strength, discs have an average M & E (mechanical and electrical) strength. Failure of a longrod is seen by mechanical parting under tension; M & E failure is loss of electric strength between cap and pin caused by cracking of the buried ceramic. Disc insulators do not part, physically, when the ceramic fails, but retain much of their initial strength by locking of the stub within the cap.

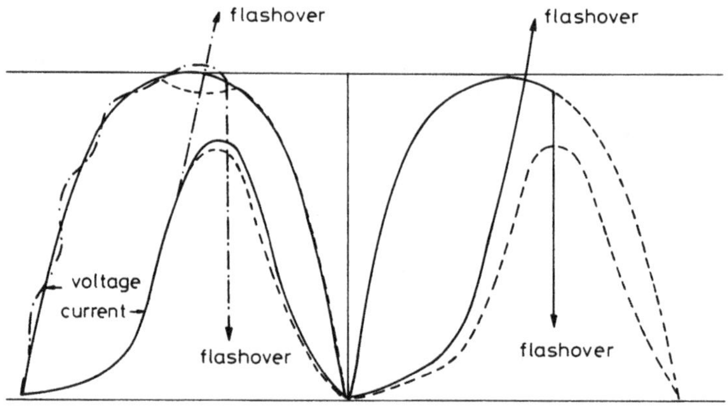

Fig. 12.9 *Interactions between source and flashover development*
2nd half-cycle inverted, for clarity
(*a*) —— Negligible source impedance: current goes to flashover
(*b*) --- Regulation reduces voltage, inhibits flashover
(*c*) --- 'Ringing' overrides regulation

If permanence of mechanical properties may be assumed, all that is needed is to measure the average strength, the standard deviation of this strength and the proof-test load. The assumption is implicit that excursions of load which do not exceed the proof value, vibration, cycling both thermal and mechanical, weathering and all other hazards of service life on the line, have no effect on the mechanical properties. In fact, ceramic insulators do change with time, but may generally be taken to have safe lives exceeding those of the other components of the line.

For fibrous-cored composite insulators, assumptions of invariance cannot be made, as stated in Section 2.4. Loss of strength under load is known to occur in ideal conditions linearly with logarithm of time. The slope is temperature-dependent and is also affected by other variables of environment and load. Many testing philosophies have been proposed for dealing with these progressive deteriorations.

In the USA some 17 tests are under consideration, including vibration, torque, stiffness, deformation under lifting and proof tensions of different severity. In Europe the International Electrotechnical Commission is in the process of standardising tests which would cover the instantaneous failing load and determine the slope of the time-load line, by exposures for about 100 hours to 75% of the average failing load.

There are good reasons for caution before accepting the validity of relatively simple tests as predictors of mechanical behaviour in composite insulators. One is the difficulty in defining failure: catastrophic mechanical collapse does not occur in fibrous composites as it does in brittle ceramics. In a core containing several millions of fibres, how many must part or separate from their matrix before constituting a 'failure', and how can such events be monitored? One cannot even assume a relationship between creep, which is easily measured, and proximity to failure, in real insulators, since it is almost always stress concentrations, at or close to terminations, which are the fatal agents.

Routine proof loading of some kinds of composite insulator has been a common practice for many years. It may be that the addition of non-destructive supplementary tests, such as observation of ultrasound emission under varying load, may find application in assessment of incipient damage.

Chapter 13

Conclusions from pollution tests on insulators

13.1 Scope of Chapter

In this Chapter we shall first consider the extent of agreement between the principal artificial tests and natural tests, then cite some general rules or laws which have been deduced from the various test programmes: these parts relate to the electrical performance. We shall then deal with the lifetime of insulators, citing the main outcome of tests on retention of surface properties, especially for glass, polymeric and resistive glazed insulators, and on deterioration leading to loss of mechanical factor of safety. Finally we shall deal with the essential matter of the value of pollution testing: how good is the agreement between results of tests, predictions from tests and behaviour of insulators on real systems in the field?

13.1.1 Agreement between artificial and natural test results

On salt-fog testing, it has already been stated that the system which is now standardised in IEC 507 was originally based on results from outdoor tests with selected insulator types. We now look at the extent of agreement between salt-fog results and those from outdoor tests, mainly in the tripartite collaboration between England, France and Italy, on a wide variety of different insulators.

Some 23 types of insulator were tested, at Brighton in England, Martigues in France and Santa Caterina in Italy, and the results compared with artificial salt-fog test data. Reasonable agreement was obtained between time to flashover in natural pollution and withstand salinity in artificial pollution[114] (Fig. 13.1). These included both disc and other types of insulator. A comparison was also made in England between results from 400 kV substation insulators tested, respectively, in salt fog and the Brighton Testing Station. None of these was in the original CIGRE set on which the salt-fog test was based and all of them were of very large size, in comparison with the CIGRE insulators.

This major study by Ely, Kingston and Lambeth[75] compared both withstand-salinity values at working voltage of 240 kV, and flashover-voltage values at fixed severity of pollution (80 kg/m^3), with results from natural pollution. Practical difficulties arose in the latter case because not all the insulators flashed over in the test period, requiring some method of comparison based on surge counts to be applied. Only three out of a total of 28 comparisons did not then support agreement between artificially and naturally obtained orders of merit, one of these of low significance.

Although this result gives impressive support to the validity of the salt-fog test, it must be seen in the light of recent knowledge about heavy-wetting flashover. Some of the flashovers at Brighton may have been caused by torrential rain whereas all the salt-fog tests were in the 'light-wetting' condition, for which a different order of merit prevails.

It will be recalled that the basic order of merit for the CIGRE insulators was established in various types of pollution, including industrial, whereas the collaborative tests and the Brighton tests were made in marine pollution. Less data exist to support a validation of the salt-fog test against natural pollution of other types than marine. Tests in industrial pollution at Friesenheimer Insel, near Mannheim, by Verma *et al.*[190] agreed with the salt-fog test in assigning better performance to an anti-fog string than to one longrod, in the order of merit, but disagreed in the relative placings of other longrods. In the latter case, however, the form of the profile which performed well outdoors is such that it must be vulnerable to any degree of wetting capable of giving drips which span the sheds. The result suggests that the real pollution is not severe and the wetting conditions are anomalous, at Friesenheimer Insel (Figure 13.2).

Although the data from desert-type pollution is sparse, work done in Bahrein, by Lambeth, Kuipers and Jumah[191], did show agreement with the salt-fog test ranking of 33 kV line insulators.

On the second artificial test which is standardised in IEC 507, the Kieselguhr test, data again come from the work of Verma, previously cited. Collaborative work in three German laboratories led to the conclusions that the solid-layer test distinguishes only those insulators whose performance in natural pollution is obviously different, and that the test requires simplification. On other pre-applied artificial tests, such as the Japanese methods, good agreement with natural test results is claimed, but discrepancies appear to arise between those results and others obtained elsewhere, on such matters as the effect of insulator diameter and linearity of performance with length. A clue to these discrepancies is to be found in Lambeth's comparison of standard deviations for the various methods (Table 12.2).

13.1.2 Laws of behaviour of polluted insulators

Some eight laws or rules of insulator behaviour have been established by the various tests[114,121]. These are listed below and followed by illustrative comment, where necessary:

(*a*) The performance of an insulator depends on its attitude. In general, but not always, inclination from the vertical improves performance.

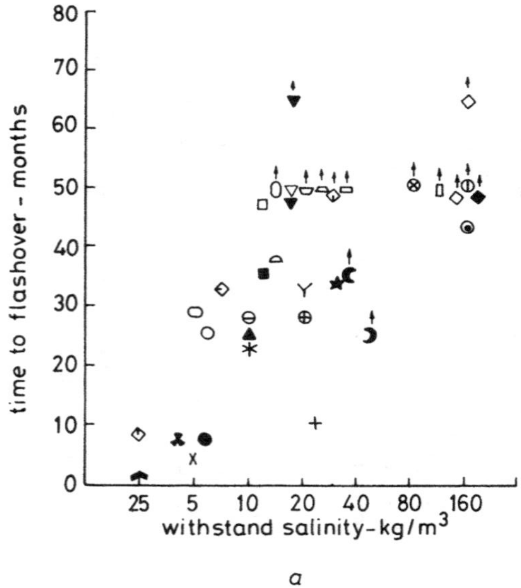

ins type	no. of units	configur(1)	test voltage (kV)	Brighton	Martiques	S. Caterina
A	3	v	84			◇
	5	v	84			◇
	7	v	84			◇
	9	v	84	◆		◇
B	10	v	84	▲		
C	8	v	84	▼		▽
D	1	v	84	●		○
E	1	v	84	■		□
F	1	v	84			◠
G	7	v	84			◯
	9	v	84			0
H	7	v	84			▭
	9	v	84			▯
I	14	v	120		⊖	
J	14	v	120		⦶	
K	11	v	120		⊙	
L	1	v	120		⊕	
M	1	v	120		⊗	
N	3	v	240	★		
O	3	v	240	+		
P	1	v	240	✱		
Q	1	v	240	x		
R	1	v	240	☽		
S	1	v	240	Y		
T	8	i	84			⏢
	8	h	84			⏢
U	1	v	240	⌃		
V	1	v	240	☘		
W	1	v	240	☾		

(1) v = vertical
i = inclined
h = horizontal

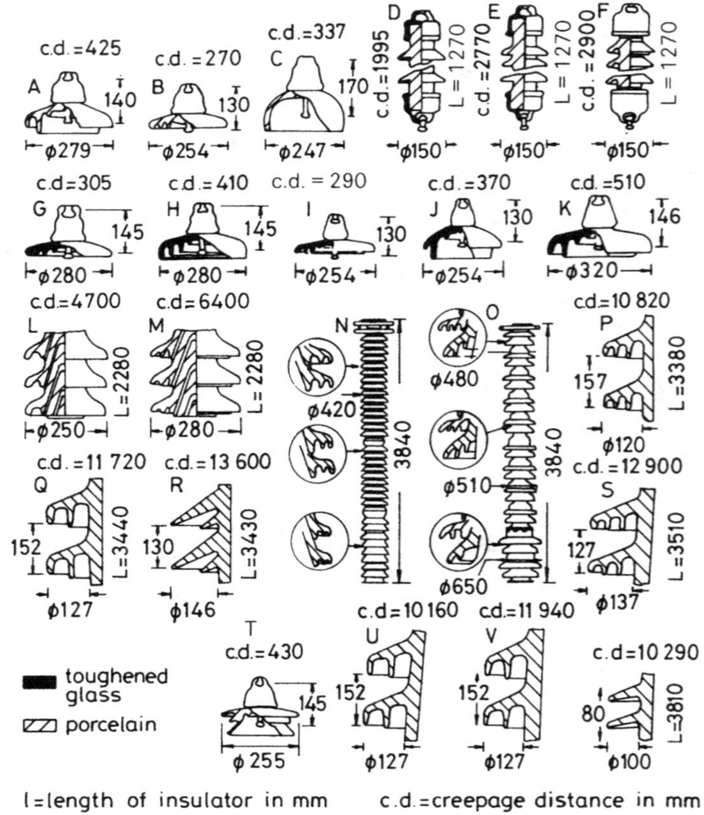

Fig. 13.1 *Validation of the salt-fog test*
a Times to flashover in natural pollution related to withstand salinity.
b Tested types

(*b*) In duplex sets, i.e. pairs of strings with the discs closely spaced, there is no electrical deterioration of performance arising from interaction although there may be aerodynamic effects. The use of V sets or inverted-V or crossed-X sets instead of duplex sets is always electrically beneficial.

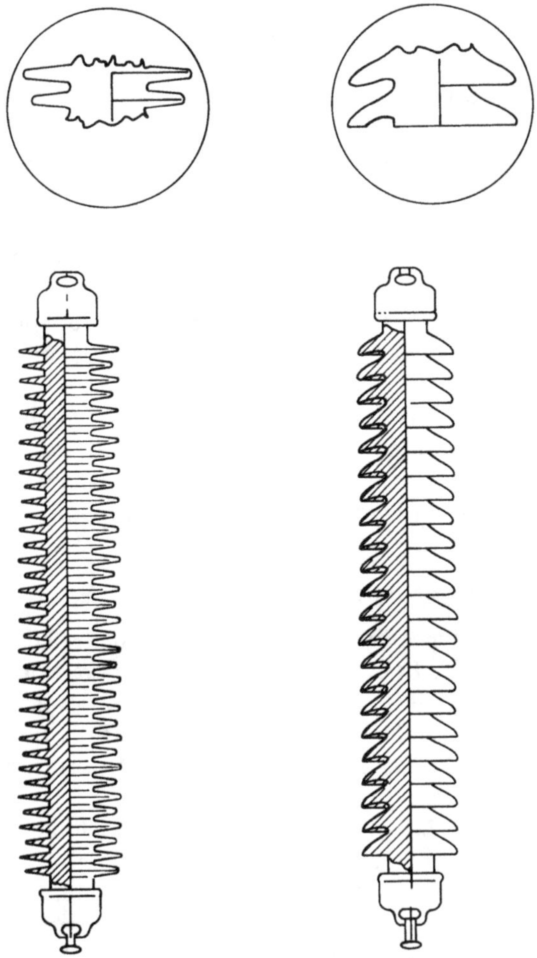

a *b*

Fig. 13.2 *Pollution severity is not a single parameter*
Creepage/axial length: (*a*) = 3·35; (*b*) 2·38
Shape (*a*), long creepage, poor profile, out-performed shape (*b*) in natural pollution
The salt-fog ratings were reversed (see also Fig. 10.7)

(*c*) If a string of disc insulators is extended by addition of units the electrical performance rises in direct proportion, at least up to 750 kV system voltage. There is controversy about linearity at higher voltages. Linearity has been

established for several types of substation insulator, including large barrels, up to lengths for 400 kV systems: again there is controversy at higher voltages and for some types of insulator having large internal capacitances.
(*d*) There is proportionality between flashover electric stress and withstand salinity. The slope of the function (flashover voltage against severity) is not the same for all insulators; important differences exist between plain shedded and highly convoluted substation insulators.
(*e*) For a given category of shape and surface, flashover voltage increases with creepage length.
(*f*) The flashover voltage of post and rod insulators falls with increasing core diameter, up to some 0·35–0·40 m, above which it levels off. Japanese results are discrepant here, showing continuing falls up to 50% over the range 0·1–0·5 m. Taper has no effect on performance, except as mentioned below under heavy-wetting effects; again, Japanese work is discrepant.
(*g*) In general, internal stress control has no effect on pollution flashover. Exceptions have been claimed for large bushings and for some highly capacitive assemblies.
(*h*) Identical shapes of insulator behave identically whether in porcelain or glass, so long as the insulators have undamaged surfaces. There are strong differences from the performances of polymers, especially when weathered, and from insulators whose surfaces have been treated or coated.

Detailed comments are as follows:

(*a*) *Attitude*: Several types of insulator string were tested upright and with inclinations[121] up to 90°. In every case marked improvements resulted, corresponding to increases in effective length up to 20%, from vertical angles as small as 5°. The effect was complete at 20°; i.e. there was little further improvement between 20° and 90°. Similar effects were found with some, but not all, of the posts and rods which were tested.

There are important exceptions to the rule that inclination is beneficial, all apparently related to draining. Some large-diameter housings perform worse when horizontal than vertical. Strings of heavily convoluted anti-fog discs may behave similarly. It seems likely that droplet motion causes flashover, in these cases. Large tapered bushings, when mounted at an angle which partly cancels the taper, leave a near-vertical edge down which severe cascading may occur. Special treatment, by Booster sheds or creepage extenders, has cured some of these ills (see Figs 14.3 and 17.5).
(*b*) *Duplex sets*: Parallel strings of insulators, both cap-and-pin and longrod, are a common feature of 400 kV lines which use multiple subconductors. Sets of four strings are common in England, in tension positions at angle towers, while sets of six have been used for some long river crossings.

It was feared that, because of random occurrences of dry banding on different strings, there would be reduction in flashover voltage associated with bridging sparks between strings. The international collaborative investigation ruled out this possibility, in joint tests[121].

The benefit of using V, Λ or X sets of units[175] arises from elimination of aerodynamic interactions between strings by good spacing, inclination of the strings and the possibility of introducing additional units into a given clearance, from crossarm to conductor.

(*c*) *Linearity*: From the simple theory of flashover (Chapter 11) it would be expected that either addition of units to a string or extension of a post or column, with dielectric of the same geometrical form as that of the original, would produce a proportionate increase in flashover voltage for a given severity of pollution. There are known large variations of stress along an insulator in the field, arising from displacement current in the stray capacitances to line and ground when the string is relatively unpolluted. When it is lightly polluted, other variations arise from the leakage resistances which now shunt the internal capacitances. Once flashover is approached, however, the surface conductivity swamps electrostatic effects: leakage currents are now of the order of 10^{-1} A against the capacitive currents no higher than 10^{-4} A, at least for discs. Many tests have been made on strings[192,185,186] all of which, with the exception of those by Kawai, confirm linearity, certainly up to lengths of 6 m.

An investigation of linearity was made by Ely and Lambeth[75] on large barrel-type insulators. For such shapes there are not only possible electrostatic perturbations of the surface electric gradient, but also obvious differences between top and bottom sections caused by accumulation of collected water and cascading under bombardment with artificial fog. The mean flashover stress was found to be close to independent of length (Table 13.1).

Although linearity can safely be assumed as a guide to selection of insulator lengths, and is most valuable where data already exist from low-voltage lines in areas where higher voltages are to be introduced, it must not be confused with the relation between flashover voltage and creepage length. If all that is done is to pack more creepage into an existing clearance, by change of shape of insulator, a proportional improvement will not necessarily be obtained in performance (see (*e*), below).

(*d*) *Flashover voltage or withstand salinity as index of merit*: In general, either flashover voltage in a given salinity of fog, or withstand salinity at a specified – usually the working – voltage, are equally valid indices of merit and linear proportionality would be expected between them. In particular cases, however, departures from the rule occur. Ely *et al.* showed[166] that plain shedded barrel insulators were discrepant from other shapes, especially anti-fog profiles with multiple skirts. The consequence was an apparent reversal in order of merit, depending on the ambient severity of pollution, between plain shedded and antifog shapes: (Fig. 13.3).

(*e*) *Effect of creepage path length*: We consider the practical case of a pre-determined clearance between live point and ground into which insulation has to be packed for maximum effect. For insulators of a given type, say cap-and-pin discs

Table 13.1 *Linearity of flashover stress: Barrel insulators*

Insulator	Section tested	Leakage path (m)	Mean FOV (kV)	Mean FO stress (kV/mLPL*)	Standard deviation (kV/mLPL*)
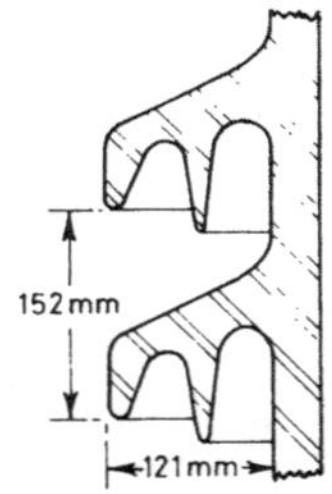 (parallel barrel)	All 22 sheds	11·02	208	18·9	0·78
	Top 15	7·54	148	19·6	0·40
	Top 8	4.05	76	18·8	0·62
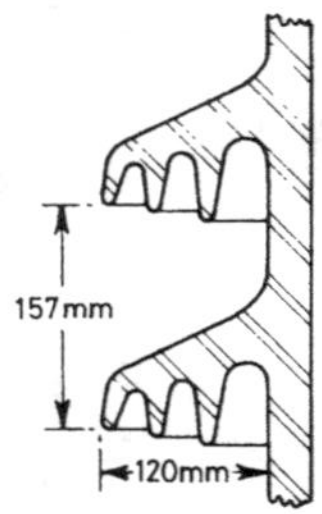 (tapered barrel)	All	10.82	203	18·7	0·70
	Bottom section	3·42	60	17·5	
	Tapered top	7·41	136	18·3	0·65
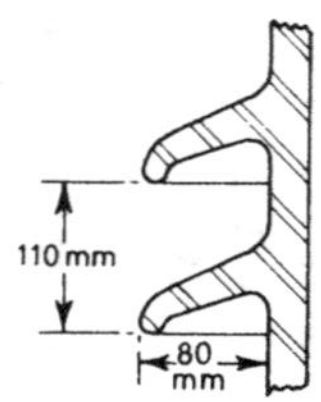 (2-unit parallel barrel)	All	7·95	169	21·3	0·34
	Bottom unit	4·84	105	21·7	0·62
	Top unit	3·11	69	22·2	0·45

*Kilovolts per metre of leakage path length: RMS values. All tests at salinity = 80 kg/m^3.

of glass or porcelain, the performance as measured by withstand salinity improves as the creepage length is increased, provided that the profile remains broadly the same. When therefore a string of 10 standard discs is replaced by a string of nine anti-fog, equal in length but more deeply convoluted, a linear improvement results (Fig. 13.1).* Major changes in shape, like adoption of aerofoil profiles which differ in both aerodynamics and draining from anti-fog, usually lead to departure from linearity with creepage length.

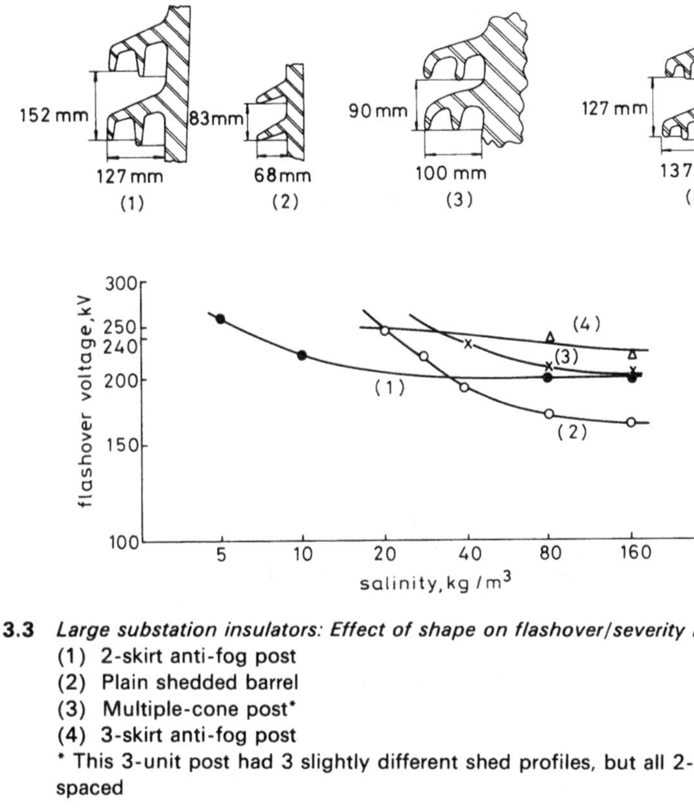

Fig. 13.3 *Large substation insulators: Effect of shape on flashover/severity relationship*
(1) 2-skirt anti-fog post
(2) Plain shedded barrel
(3) Multiple-cone post*
(4) 3-skirt anti-fog post
* This 3-unit post had 3 slightly different shed profiles, but all 2-skirt and close-spaced

For longrods of equal length, performance usually rises with creepage. Generally, worse performances than those of anti-fog strings are observed. Very close-packed skirts, especially where drip rings are omitted, give[193] poor results in salt fog but good ones in pre-applied pollution tests. Motor insulators, in some ways intermediate between discs and longrods, show direct improvements with increasing creepage, even when this is obtained by polymeric additions, whereas longrods are more affected by shape (Fig. 13.4).

Large substation insulators, like barrels and posts, depart widely from linearity of performance with creepage. When Ely and Lambeth's data from 400 kV

* Very recent work, in England, shows that this is not always true.

substation insulators[166] are plotted, either with flashover voltage or withstand salinity against creepage, little correlation appears. Some dependence does exist between flashover stress and density of creepage, i.e. when effective height is taken into account (Fig. 13.5). Their principal discovery, that there are strong changes in order of merit with severity of pollution, especially between plain and convoluted shapes, obviously undermines the general policy of specifying substation insulation solely by creepage. There are interesting similarities with longrod behaviour, in this respect.

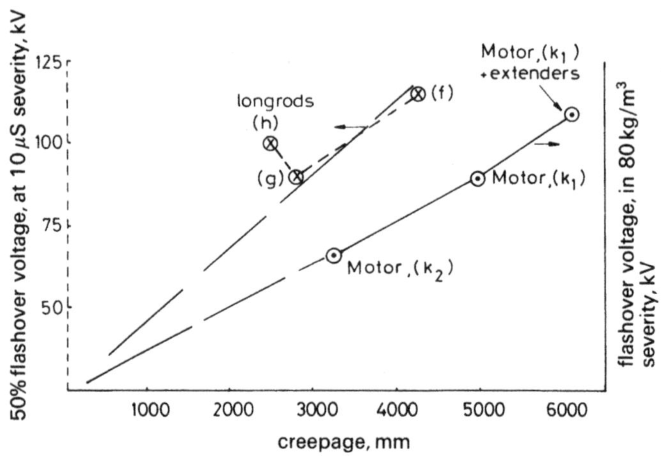

Fig. 13.4 *Variation in performance with creepage length: Motor and longrod types*
Motor types ⊙ perform linearly in salt fog, although shapes differ widely and one has polymeric creepage extenders. See Fig. 13.6 for shapes.
Longrods ⊗ depart from linearity, when tested with preapplied pollution.
(Data from Reference 190 and FGH tests, courtesy of Raychem UK, Ltd.)

Draining effects must operate with large insulators like these. Ely and Looms[35] showed the bad effects of large horizontal areas of skirt, causing slow flow of pollutant. Experiments with horizontal posts also show[194] that direction as well as length of creepage is relevant. If h is the horizontal part of the total creepage L, flashover voltage is roughly proportional to $L - (h/4)$.

The importance of draining effects is confirmed by tests with pre-applied pollution on vertical and horizontal insulators of similar type. No difference is found[187] if the surface conductivities are the same in both cases.

(*f*) *Diameter effects*: It might be thought that flashover voltage would decline indefinitely with increasing diameter of insulator because the overall resistance, which limits the arc current, would also decline similarly. However, it has been shown (Section 11.3) that more than one arc can burn across a single dry band, provided that sufficient peripheral resistance exists between their roots. Erler[152] treated the surface as divisible into strips parallel to the axis, each of width about $3L/2N$, where L is the creepage length and N the number of dry bands.

Although the general view is that the fall in performance is complete at some

0·35 m diameter, interesting effects occur at the other end of the scale, where diameters are very small. Ely, Looms and New showed that extraordinarily high flashover voltages were obtainable from polluted monofilaments of diameter about 2–3 mm, evidently because of high resistance arising from small periphery[195].

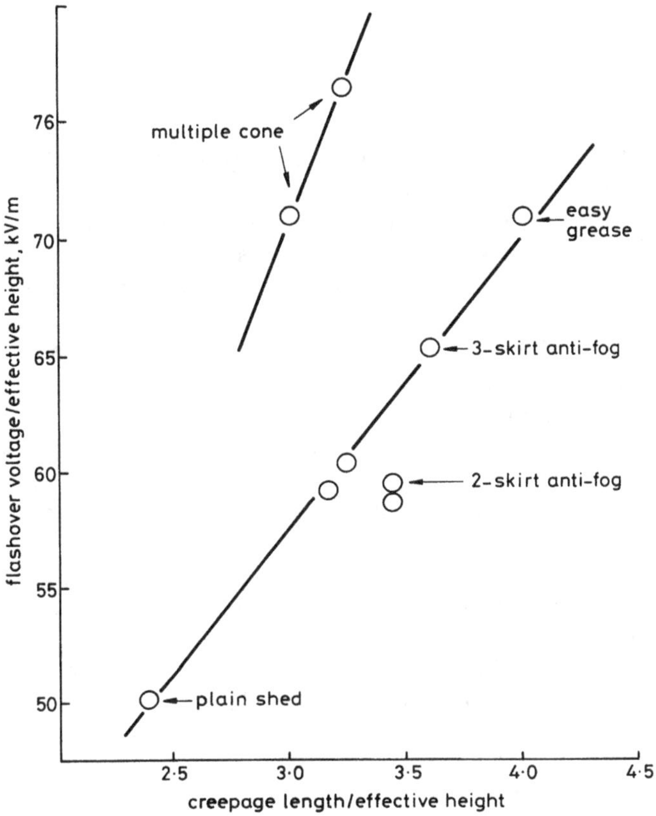

Fig. 13.5 *Dependence of flashover stress on density of creepage: Large substation insulators*
For large insulators in severe pollution, (80 kg/m^3), performance improves as specific creepage length increases. Multiple cones excel.
In lower severity, plain sheds are best: different orders of merit prevail under heavy wetting.
(Recalculated from Reference 166). (For shapes, see Figs. 13.3 and 13.7)

(*g*) *Internal stress control*: It is still sometimes suggested that capacitive stress control of surface electric gradient may be used to improve the flashover performance in pollution, although this view is almost entirely discredited. One of the two relevant areas is strings of discs, where artificially high unit capacitances have been tried as a means of reducing variations of voltage from disc to disc, in which they are effective, and thus of improving flashover performance, where they are not. One type of experimental high-capacitance disc used very thin dielectric shells of glass, which more than doubled the normal unit capacitance of some 30 pF: the internal gradient was sufficient to cause cases of thermal runaway. No significant difference from normal flashover voltage in pollution was observed. The other area is housings which contain large capacitive elements like bushing shells and capacitive heads for circuit breakers.

In such cases the external pollution layer is capacitively coupled to an internal voltage-grading element, and is thus itself subject to enhanced flow of displacement current.

Consideration of the relative magnitudes of the likely displacement currents, of the order of 10^{-5}–10^{-3} A, and the surface leakage currents close to flashover, of order 10^{-1}–1 A, shows that no interaction with the propagation processes is probable. Insofar as high capacitance acts in a string of discs to make the voltages per disc uniform, it is useless, since the surface gradient is close to uniformity at flashover in any event. This is evidenced both by Hampton's experiments and by the established linearity of flashover voltage with number of discs.

Ineffectiveness of stress grading has been confirmed by several workers. Brzuska[196] used pre-applied methylcellulose to compare housings of identical form with and without stress grading, and the author's colleagues used the salt-fog-test for the same purpose[75]. Kawai[186] showed what appeared to be capacitive effects, using the kaolin and steam test; it is, however, notoriously difficult to get uniform wetting with such a test, and hence to ensure a reliable comparison.

The large arcing horns, which are used on transmission insulators for 400 kV in England, and the corona rings, which are features of HV and UHV transmission elsewhere, have significant stress-grading effects and indeed are used to abate generation of radio interference. These have been shown, again, to have no effect on flashover of the associated insulators. In recent times, however, external stress grading of this type has been found advisable on some types of polymeric longrod, both to minimise surface discharges near the live terminal and to reduce the chance of internal propagation of damaging tracks along the fibrous core[79].

(*h*) *Glass equals porcelain, in a given shape*: No significant differences have been found between the performances of glass and porcelain insulators of identical shape in equal pollution levels, under AC. Differences are[197] alleged under DC. There is no doubt, however, that surface changes occur with time under stress in both materials, which influence the behaviour of the insulator. In addition to the large variability in surface resistivity with changes in humidity, already mentioned and common to both, there are short-term conditioning effects, whereby insulators under test show a decline in flashover voltage to a stable value, over about the first hour of subjection to heavy surface discharges in contamination by fog, and long-term increases in surface leakage caused by deterioration of the surface. Neither effect is well understood.

Conditioning is sometimes attributed to the removal of surface films of grease, and stringent cleansing requirements form part of many artificial test routines, for this reason. Some sort of surface activation also seems a probable contributor, however, since insulators usually revert to their unconditioned state after a rest period in circumstances where re-contamination is remotely unlikely. Some polymeric insulators, under test, exhibit short- and medium-term conditioning effects which almost certainly arise from chemical activation of this kind[197].

Surface deterioration is easily visible on glass insulators which have suffered attack by discharges, especially in heavy salt pollution (Fig. 2.10). In similar conditions the damage to porcelain is relatively slight; for both materials it seems likely that there will be a resultant increase in surface leakage. The decline in flashover voltage of old insulators has been attributed to this effect[166].

13.1.3 Influence of shape on insulator performance

13.1.3.1 Cap-and-pin strings, longrods and line posts: Although much more work has been put into testing line insulators than any other class, the results are not yet altogether free from ambiguity. Current work will, it is hoped, resolve two questions, respectively the true value of aerofoil shaping and whether or not desert-type pollution behaves differently from industrial and marine. For *cap-and-pin strings* the main facts are as follows; initials refer to Fig. 13.6.

Where convoluted skirts are used (*a*, *b*, *c*, *d*), performance does not vary much with shape for a given creepage length. Shapes with a dominant long skirt (*b*) perform marginally better than those with near-equal skirts (*c*). In shapes with several long skirts, care is needed to avoid designs in which two skirts terminate in a plane: aerodynamic resonance effects have caused acoustic noise nuisance with shapes like (*d*) (Section 16.4).

Desert discs with open aerofoil profiles perform poorly in the salt-fog test, but significantly better, in natural tests and service, than would be expected from their creepage lengths. Not much quantitative data are yet available, but indications are that shape (*e*) is at least as good as (*b*), even though its creepage is some 30% less. It must be realised that aerofoil discs are likely to have much greater diameters than anti-fog for comparable performance, which may complicate the design of duplex and V sets (Fig. 13.6 *d*, *e*).

These remarks apply to vertical strings and to discs of about equal mechanical rating. In general, electrical performance falls as rating rises: this may arise from the large metalwork and worse aerodynamic shape. For horizontal strings the effects of disc shape are less firmly established. It seems that small diameter and minimal convolution, both promoting rapid draining, are beneficial. In any case there is less economic incentive to push the performance of horizontal strings to the limit than with vertical: to introduce additional units into a horizontal string, the designer has no need to modify the tower geometry. Very long strings of very small units can thus be used: such designs have performed well in combined desert and coastal pollution in North Africa.

For *longrods* and *line posts* the shape is dominated by the need to provide adequate core section for the mechanical loads (*f*, *g*, *h*): in general, poorer electrical performances at a given creepage length are obtained than for cap-and-pin strings, at least in Italian and English coastal test sites. Thus, longrod (*h*) has about the same creepage as a 7-unit string of anti-fog disc (*b*), which easily out-performs the longrod although some 30% shorter. Some shapes of rod having close-packed sheds, either helical or near-planar, have performed much worse in severe pollution than would have been expected from their long creepages. The special claims, of good washing in rain, which have been made for the helical shape are apparently illusory (*i*, *j*).

In longrods and line posts, the absence of intermediate metal to act as arc stopper adversely affects their performance. Kjølby[198] made replicas of cap-and-pin profiles, including the metal caps, in all-porcelain: these evidently had more creepage than their originals, since metal was replaced by porcelain at all intermediate points. Their performance, at the Brighton Testing Station, was significantly worse than that of the metal-carrying string. Although this result indicates a possible advantage of a string of Motor insulators (*k*) over a

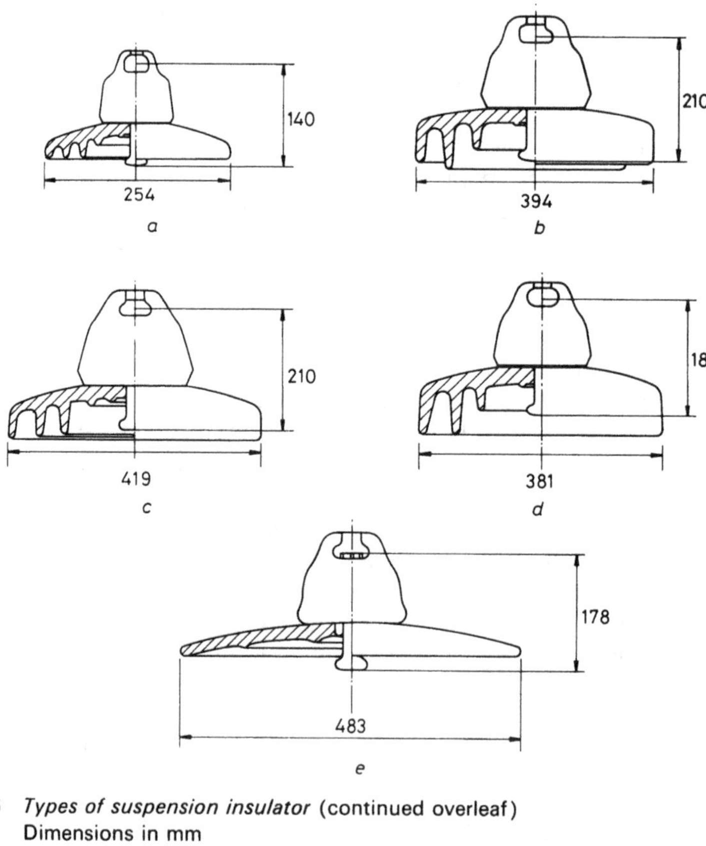

Fig. 13.6 *Types of suspension insulator* (continued overleaf)
Dimensions in mm

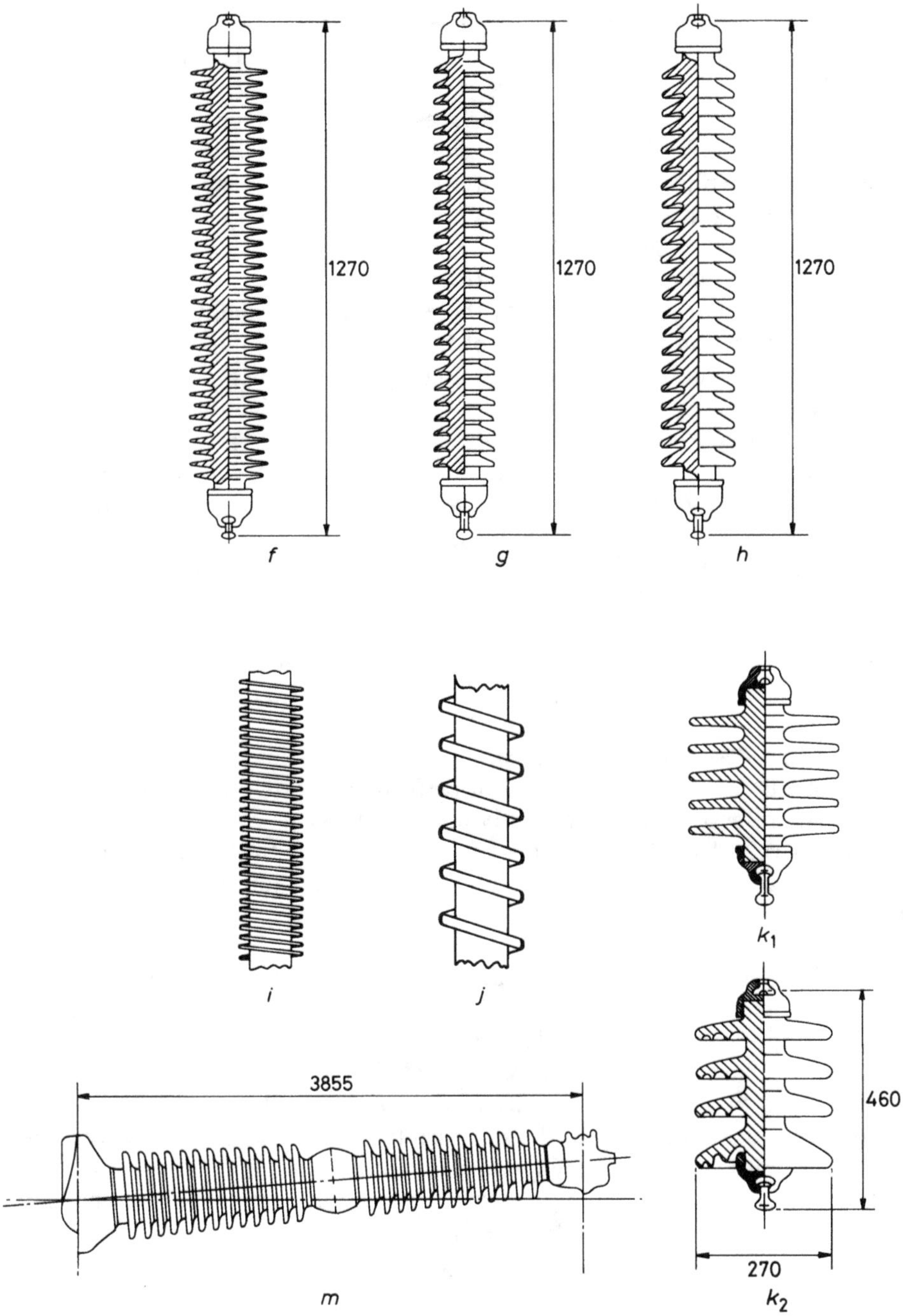
1270
1270
1270
f
g
h
i
j
k_1
3855
460
270
m
k_2

comparable longrod, much axial clearance is wasted by the couplers in a string of Motor short-rods, and poor performances have resulted, for example, in Israel[199].

Combining inclination of shed surfaces with long creepage is difficult in longrods, and poor draining thus often results. All longrods share, to some extent, the known liability to heavy-wetting flashover of insulators with close-packed sheds[193], and are likely to derive benefit from ancillary water ejectors and droplet barriers (see Section 14.3.1).

These comments, like those on strings, relate to the vertical attitude of mounting. For horizontal longrods there is evidence (Chapter 6) that some degree of protection against impact of level-moving pollutant is beneficial, as well as minimal interference with draining. Slightly inclined planar sheds were found to perform better than those at 90° to the axis[48], when sideways-bombarded with pollutant in the salt-fog test.

The relationship between orientation of horizontal or inclined insulators, position of main source of pollution, and electrical performance is far from clear. It might be thought that to point the underside of a string or rod, where most of the creepage resides, towards the main source of contaminant – such as the sea, in marine test sites – would obviously increase the flashover rate, but this is not found to be universally true. Significantly better results have sometimes been found than for the converse mounting position. A possible contributor is the 'aerodynamic bubble' which is found, in some cavities, in wind tunnel work[104] (see also Section 11.3).

Line posts, designed to work in cantilever and sometimes using tapered cores (*m*), generally do not compare well in performance with strings or even longrods, and are best confined to lightly polluted areas.

Various rules have been proposed for dimensioning the profiles of longrods and posts; broadly similar to those applied to substation insulators, they are dealt with below.

13.1.3.2 Substation insulators: The conclusions of Ely and Lambeth's study of substation insulators[75], insofar as shape is concerned, were that most types of 400 kV support or housing insulator perform badly, in comparison with types for overhead line duty, only half of those tested having been found capable of withstanding two years without flashover in severe pollution (Brighton); that large barrels were the worst of the standard types; and that pedestal posts were poor, per unit height, compared with many others.

For barrels, the descending order of merit, for a given height and core diameter, was as follows (see Fig. 13.7):

re-entrant conical	(A): also called 'easy grease'
3-skirt anti-fog	(B)
2-skirt anti-fog	(C) equal rating with (D)
3-shallow-skirt	(D)
wide-spaced plain	(E)

The other shape variables, diameter and taper, were found not to have significant effects, but this conclusion is now known to be subject to correction with inclined insulators.

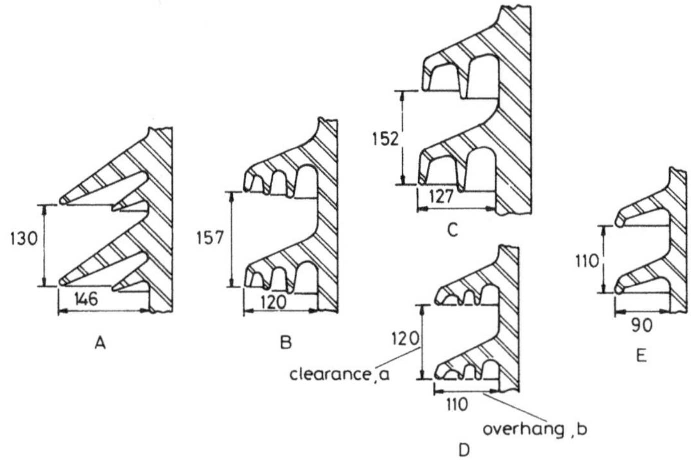

Fig. 13.7 *Profiles of large substation barrels*
Dimensions in mm
Descending order of merit, left to right
A Easy grease: B 3-skirt anti-fog: C 2-skirt anti-fog: D Italian 3-skirt anti-fog: E Plain shed

Other work, generally with pre-applied pollution, has led to rules for profile design, especially concerning optimum ratios of clearance a to overhang b (Fig. 13.7). These are claimed to apply both to substation and line insulators. Steyer[200] suggests that a/b should be about 0·8, while Pohl[201] proposes limits of 0·6–1·5. It is interesting to note that none of the insulators of high merit, mentioned above, obeys Steyer's criterion, and that all fall within Pohl's limits!

Two other rules, derived partly from tests and partly from research, are also applied to profile design in some countries. These are, respectively, the J-number and P-number criteria. The J-number, well described by Robinson[202] (Fig. 13.8), relates to the avoidance of inter-shed flashover. The probability of air breakdown, or of droplet-initiated breakdown, increases as the point-to-point surface voltage, given by the product of surface-resistance/unit-length and leakage current, rises. Robinson gives a J/l ratio not less than 0·5 as desirable for heavy pollution, where J is the clearance and l the length.

The P-number, a concept developed in Australia with the severe pollution of the Latrobe Valley as the target for remedial action, takes account of several shape factors including average diameter, creepage length, shed spacing and insulator length, as well as the fraction of creepage which is protected

$$P = 11{\cdot}7\,\frac{AB}{C}\,\frac{L_a^2 L_s}{V_+^3}.$$

Here V_+ is the system voltage (kV), L_a, L_s the axial and creepage lengths (cm) of which L_{sp} is protected. A is a constant given by $(1{\cdot}5 - 0{\cdot}15\, b/a)$, where b, a are average shed length and spacing, for cylindrical and post insulators. For cap-and-pin discs and pedestal posts, A is respectively equal to unity and to 0·9. B is given by $1 - L_{sp}/2L_s$ and $C = 1{\cdot}2, 1{\cdot}0, 0{\cdot}8$, respectively, for $V_+ = 66, 132, 330$ kV.

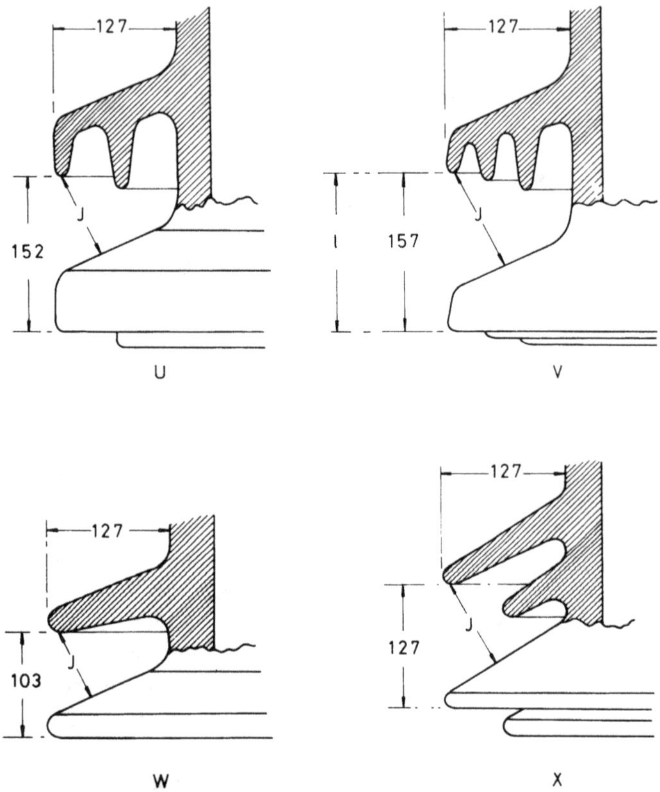

Fig. 13.8 *Robinson's J-number*
Dimensions in mm

Insulator	Creepage/length	*J*/length, *l*
U. 2-skirt AF	3·50	0·58
V. 3-skirt AF	3·50	0·70
W. Close plain	3·10	0·73
X. Re-entrant cone	3·85	0·76

P is then required to equal or exceed the critical severity number P_c, for the area where the insulator is to work (Table 13.2).

An attempt is in progress, at the time of writing, to produce a guide to the selection of insulators. This task has been assumed by the International Electrotechnical Commission, Technical Committee 36. The guide will, apparently, embody empirical rules of the sort already mentioned above, but will not give supporting research data for its various assertions. The value of such a document must be questionable: it would give spurious authority to unproven propositions and encourage the incorporation of folklore into purchasers' specifications.

Table 13.2 *Critical pollution numbers, P_c: Australian system**

Severity class	Max. ESDD (mg/cm^2)	Critical P_c	Indications of severity
A = very light	0·10	10	More than 16 km from coast No other pollution source
B = light	0·14	14	8–16 km from coast Other sources, light
C = moderate	0·20	20	3–8 km from coast Other sources, moderate
D = heavy	0·30	30	1·6–3 km from coast 0·8 km from salt lake, river Other sources, heavy
E = extreme	0·45 and higher	45	Within 1·6 km of coast Other sources, severe

*Classification after Hurley and Limbourn[135]
Strict validity claimed only for conditions in New South Wales

The main elements are understood to be as follows:

A *minimum distance* between sheds will be suggested, about 30 mm minimum, to avoid bridging by rainfall. (Without relationship to shed collecting area, such a fixed spacing seems arbitrary.)
A *spacing/overhang ratio* will be proposed, apparently following Steyer's criterion of 0·8.
A *creepage/clearance ratio*, relating the creepage path between two points and the air clearance between them, will be set at 5 as maximum.
A difference between *consecutive overhangs* of 15 mm will be proposed.
Shed *inclination* will be set at 5° minimum (top) and 2° (underside).
A *profile factor* will be suggested which will relate the design of the insulator to the severity of pollution, and will apparently follow to some extent the philosophy underlying the Australian *P*-number, in involving both creepage length and arcing distance.

It does not appear that this type of documentation is receiving much support from countries in which insulator research is most active, and where it may well be felt that Canute was not the last man to attempt to override the inconvenient laws of physics by edict. While awaiting the outcome with interest, many researchers in the field of insulators will prefer to investigate and innovate rather than accept and codify.

At the time of writing the status of research into the behaviour of outdoor insulation is somewhat uncertain. Important steps forward in understanding over the last two decades have been followed by doubts, especially in the areas of very high voltages, DC and unfamiliar types of pollution, about the very fundamentals of the flashover process. These are not well founded[203]. A proposal which has surfaced intermittently over the post-war period, that insulators could be designed by computer, seems unpromising: as recently as 1983 the computer-aided design of profiles was confined to the 'zero pollution', i.e. trivial, case[204].

13.2 Deterioration: Test results

13.2.1 Scope of tests

Tests on polymeric components and other materials have already been covered in Section 6.5. These give approximate guidance on prospective lifetimes of different systems. However, artificial tests using powerful sources of voltage, and outdoor tests giving synergy between weathering, pollution and current, have thrown up some unforeseen results[205].

Since, also, it is difficult to simulate all the hazards of real service, such as mechanical load cycling, overvoltages, vibration or twisting, some important causes of deterioration have been masked. These include corrosion and brittle fracture of fibrous composite cores, as well as damage to glass and porcelain associated with cement and buried metal.

The broad picture from all types of test taken together is that porcelain deteriorates negligibly, glass significantly and polymers rapidly: in worst cases, the respective times to serious loss of properties might be decades, years, months. Recent work suggests, however, that relative reliabilities are much less disparate when all aggressive factors are taken into account. Corrosion and impact effects have down-graded some porcelain types, avoidance of specially contaminated places has up-graded glass, and technical improvements have much up-graded some types of polymeric insulator.

13.2.1.1 Deterioration of porcelain: After artificial-fog tests lasting hours, merely trivial surface changes are visible on porcelain insulators except where discharges have been confined close to the surface under polymeric additions (Chapter 14). Slight burn damage is sometimes seen near the terminals, if the source is a very powerful one.

Repeated flashovers under switching surges of durations only a few milliseconds, but carrying heavy transient currents, have produced surface roughening, but only under some electrical conditions of polarity and on certain designs, especially the multiple-cone posts. This test has also dislodged part of the pin-cavity sealant in some large cap-and-pin discs.

In outdoor tests on insulators coated with protective greases or silicone pastes, where the test duration has been long enough to lead to contamination of the greases or loss of oil from the pastes, visible glaze damage has resulted (Fig. 13.9). None of these cases has seriously disabled the insulator.

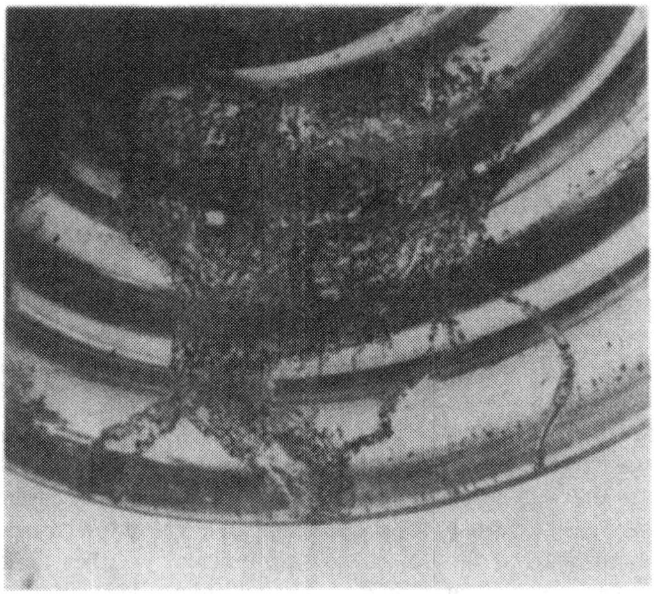

Fig. 13.9 *Discharge damage caused by thin silicone layer (after Reference 175)*
Glass is more severely attacked than glazed porcelain

Resistive glazes used to improve electrical performance (Chapter 14) have all deteriorated under outdoor test. Those on cap-and-pin discs have shown major electrolytic and discharge attack, especially in regions of high leakage-current density around the pin cavity, and some discs have cracked. Cracking has been ascribed either to gross erosion – damaged regions of glaze may carry permanent arcs – or to rusting in the pin cavity, aggravated by the standing current. Post and barrel types have also suffered visible spalling of resistive glazes, leading to local bands of very high resistivity. Few of these have failed, however, and all have continued to out-perform normal glazed controls, both from the aspects of leakage surge frequency and incidence of flashover. Swift[205] has calculated that displacement current flow in the bodies of such 'burned out' insulators may well be sufficient to maintain some degree of stabilisation.

13.2.1.2 Deterioration of glass: Glass insulators may be expected to deteriorate more rapidly than porcelain, for reasons stated in Chapter 2. However, the inferiority is to some extent exaggerated by the nature of the material. A degree of 'frosting', which will transform transparent into translucent glass, may well be imperceptible if occurring on opaque porcelain.

Artificial tests, both in fog and under impulse, do generally cause some marking on glass insulators. No corresponding decline in electrical performance is, however, recorded in the large-scale international collaborative publications[114,121].

In outdoor tests there is a strong dependence of serious deterioration, such as shattering of the dielectric, heavy localised erosion near the pin, or damage in the cavity, on the severity of the salt pollution and on the leakage-current amplitude and duration. At Brighton a short horizontal string of glass insulators, mounted very close to the ground, showed serious damage in one year, without, however, any deterioration of the by-passed unit, used for leakage current measurement at the earthy end of the string. Comparable insulators, at the same site but mounted at realistic heights above the ground, showed little or no deterioration in longer test periods.

The 'trapped' discharges, either under spent grease, inert blanket types of pollution layer, or polymeric additions which cause glaze damage on porcelain, are much more deleterious on glass. Attempts to modify the performance of glass discs, by locally greasing the pin-cavity region out to the first skirt, caused very severe attack at the boundary of the grease[175].

Because glass deterioration is a function of the quantity of leaked charge, and because leakage current rises rapidly with applied electric stress, failure of one or two insulator discs can, in principle, lead to an avalanche or cascade failure of the whole string. Attempts to reproduce such an effect in the laboratory were made after a multi-string failure, caused by unusual weather conditions in England, disrupted power supply on a transmission line. The experiment failed and the incident remains unexplained[206].

Summarizing the behaviour of glass under test, we may say that; in many kinds of pollution, deterioration rates are comparable with those of porcelain; in severe salt-deposit conditions, especially where ancillary binding deposits exist, deterioration rates may be unacceptable; in all cases, failure is obvious and location easy.

13.2.1.3 Deterioration of polymers: There is a wide gamut of deterioration processes when polymeric insulators are tested, either in artificial or outdoor tests, affecting both the housing and the core.

Total failure by internal or external flashover has been produced under artificial-fog tests of a few hours' duration on foam-filled fibrous tubes, used as insulators in live working (Chapter 15). The weakness of these designs, in which there is no difference between creepage and overall length and no arrangements for interrupting water cascades, has been confirmed in outdoor tests, which have led to failures in less than one week. Interesting thermal effects occur, whereby the outer surface and interface are severely charred while the body of the tube retains its integrity[48].

Artificial tests on polymeric designs intended for outdoor use show complicated conditioning effects, with the surfaces of some materials exhibiting both slow and fast changes under discharge attack[48]. Longer-term tests have been devised which combine electric stress with cyclic pollution and, sometimes, artificial weathering (Table 2.11). For intended applications where special conditions of contamination exist, such as on electric railways, reactive pollutants like iron oxide may be introduced, since these are known to cause burning and chemical effects on some surfaces[80,81] (Section 6.5).

Numerous tests of a nature specific to polymeric insulators have been either proposed or actually used[207]. These include: steep-fronted impulse applications, to explore interfacial failure as a result of other challenges to the insulator like mechanical loading or immersion in water; running of high-current arcs, typically of several kiloamperes over durations of the order of seconds, to check for flammability and objectionable carbonisation; impact or dropping tests, to assess the consequences of shock mechanical loading, applied in different ways, for the water seals and interfacial integrity. Salt-fog tests, either of low salinity and long duration[208] or following IEC 507, have also been applied, in the first case as simulated contamination and in the second to measure the loss of performance after weathering. Falls of 40% in withstand salinity have been measured after two years' weathering, with some types of housing.

Other tests of a semi-diagnostic nature have also been used, in particular to explore the state of the interface. Bradwell and Wheeler[81] have used repeated immersion cycles in water, hot to cold, followed by measurement of DC leakage at stresses of about 100 V/mm, with a 'pass' level of 0·1 μA; point-source X-ray photography has been applied to finding splits and cavities within complete insulators[35].

Outdoor testing has been performed in most conditions of contamination, using test voltages up to 750 kV and, in a few stations, mechanical cycling applied hydraulically. Surface condition has been monitored and systematic measurements made of leakage current[79].

It has been found that damage is not easy to detect, even by visual inspection at close range. Suspects are best removed for X-ray or other examination. Cases have occurred of gross surface fissures[209], apparently caused by displacement current followed by invasion of pollutants, which have been invisible through binoculars and which have caused negligible changes in leakage current. Multiple puncturing of sheds has also occurred, again with no detectable electrical consequences.

In all outdoor tests on polymeric insulators the main difficulty has been the achievement of effective acceleration. In heavy pollution but under normal electric stress, corrosion effects which depend on quantity of charge may be masked: less severe pollution but higher average humidity has been shown, for some polymers but also for some resistive glazes, to be much more damaging. Increased frequency of wetting by artificial sprays[81] has been used by Bradwell and Wheeler as an accelerator, but this system must obscure effects of internal electrostatic discharges which predominate in dry, or in drying, conditions. Overvolting the test piece has the objection that, again in dry conditions, the live-end electric gradients are very high and may well cause untypical effects at the termination and interface. There is a special difficulty with polymers in that some are specifically vulnerable to the chemical nature of the pollutants which occur at the test site, while almost all are adversely affected by overlays of dust, clay or cement, and pulverised fuel ash containing acids. Highly misleading results are likely, if results from a single site, or single artificial pollutant, are relied on.

The broad outcome of outdoor tests on deterioration is as follows:

Tracking is effectively unknown, with modern housing polymers, except where inert overlays occur.
Erosion is universal with all polymeric housings, but is unobjectionable unless localised.
Discharge erosion does decrease as leakage current is reduced, as by long creepage and good surface properties, but even discharges at milliampere level are damaging, because of the high arc gradient at low current levels.
Electrochemical erosion depends on the leakage current density and can be reduced by increasing the stem diameter at the terminals.
Invasion at terminals is a near-certain symptom of impending failure of the insulator because the fibrous core is highest stressed, mechanically and electrically, at these points.

13.3 Is testing of insulators valid and valuable?

We consider the extent of agreement between the conclusions which have been drawn from tests of the various kinds and the behaviour of insulators in real service, on lines or in substations.

As far as cap-and-pin strings are concerned, excellent agreement exists between the results of the artificial and natural testing on the one hand, and the behaviour of installations on the other. Effectively all fault incidents on existing lines and in substations have been explained in terms of test data. For new lines, where predicted insulation requirements have been based on test data, the results have generally been satisfactory, even at UHV. Exceptions exist with direct-current systems; the reasons are discussed in Chapter 15.

The rules, based upon tests, of insulator behaviour, such as maximisation of creepage length consistent with acceptable shape, linearity with voltage, properties and advantages of inclined strings, have all been borne out in practice (Fig. 13.10).

The advantages of aerofoil shapes for best use of creepage have proved greater than predicted from some tests; however, the behaviour of lines in some arid regions, where both marine and desert types of contamination arise with little natural washing, has recently been such as to provoke a retreat from aerofoil to conventional long-creepage anti-fog discs. The test ambiguities may, it seems, reflect reality.

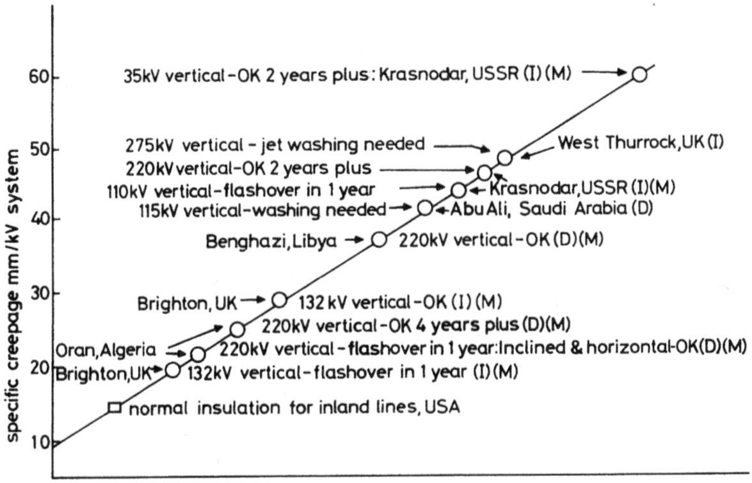

Fig. 13.10 *Practical validation of rules of insulator behaviour*
Pollution types
(I): Industrial
(M): Marine
(D): Desert

For longrods, Motor insulators and line posts, the second main group of line insulators, doubts remain. According to most tests, flashover rates with such types should, for equal overall lengths and severities of pollution, be significantly higher than for good cap-and-pin designs. In Germany, home of the longrod, there is effectively no coastal pollution and the true severities of industrial and other contamination may be lower than suspected or measured. Flashover frequencies are claimed to be no higher than with strings of discs: against this, quite high frequencies of conductor dropping are quoted[210], of 1·5 per 1000 km of line annually or even higher. Again, recent experience in arid regions has suggested that longrods are no longer favoured for the most severely polluted places and are being replaced by[191] long-creepage anti-fog designs (which, in virtually all test procedures, appear most resistant to flashover).

Line posts and Motor insulators again should, according to test data, exhibit higher than average flashover rates. Where these are in wide use, respectively in the USA and Scandinavia, their acceptability may rely on the fact that pollution levels are generally very low. It is significant that live washing is widely used in North America, where line posts are installed close to beaches and in industrial areas, and that flashover rates are high where Motor insulators are used in contaminated areas of the Levant and Arabian Gulf.

Insofar as agreement with the behaviour of substation insulators is concerned, the validity of tests is good in some subjects at least. The predicted deficiency in performance of many large supports and housings, in comparison with that of overhead-line types, has been fully borne out and has called for the widespread applications of remedies like greasing and live washing (Chapter 14). The successful diagnosis of the 'heavy wetting' flashovers, either caused by live washing or by torrential rain and in both cases affecting insulators carrying little contamination, in a variation of the salt-fog test, and the use of this test to aid development of a successful remedy for the condition, together constitute a powerful validation.

On prediction of deterioration in material rather than electrical properties, the success of testing has been limited. With the exception of the predicted cap bursting in DC insulators and the fast deterioration of resistive glazes, tests on the whole have failed to warn of corrosion-related defects in porcelain, especially cracked discs and multiple-cone posts as well as pedestal posts. The aging of glass has been overstated by tests: in service the really grave cases of attack on glass surfaces have been confined to special areas like the English Channel coast, parts of North Africa and the Western Isles of Scotland.

Testing, both indoor and outdoor, has been indispensable in the case of polymeric insulators. Without the warnings of housing erosion under weathering and stress, which tests provided, and the early proof of vulnerability of fibrous cores, the catalogue of operational failures of polymeric insulated lines would have been longer than it is.

On the economic value of testing there can be little dispute. It was estimated that the capital costs of all the test facilities, artificial and outdoor, which were used to determine the proper levels of insulation for the entire British 275 and 400 kV network, about equalled those of a 10-mile run of line. Testing costs are negligible, in comparison with those following a major error in insulator choice: the cost of replacing a failed insulator may be 100 times its purchase price for a line type, while total outage and ancillary charges arising from a busbar support failure, or flashover of a generator transformer bushing, may be several orders higher.

Interesting illustrations of predictive economies from testing include the glass insulation across the island of Laessø, part of the Conti–Skan DC scheme. This had to be selected on limited test data, some of it English, but has proved entirely successful. The last-minute warnings of vulnerability, again from tests, which governed the choice of insulators for the Italian million-volt project, also showed the cost effectiveness of the tests and of international collaboration on insulator research.

The really valuable outcome of all the insulator tests has been the demonstration that there is no 'magic insulator', no device which will have comparable effect, in reducing the dimensions of transmission apparatus, to what the solid-state device did for the computer. The rule seems inescapable: pick long strings and tall posts for dirty places.

Chapter 14

Remedies for flashover

14.1 When are remedies needed?

Remedies are called for, evidently, when the flashover frequency rises above acceptable levels. What is 'acceptable' depends on the importance of the line or substation and on the required quality of supply, in terms of outage time per year. Standards thus vary widely: for lines, a rate of about one flashover per 150 km per year is general for industrialised countries in Europe, while much higher rates are tolerated elsewhere, e.g. in rural parts of the USA. Flashovers in substations often have serious consequences, and rates lower than one per year per station would normally be called for.

The causes of flashover are part systematic and part random. In service an insulator will carry a resident layer of contamination, accumulated since installation or the last cleaning operation, which fluctuates as the resultant of depositing and purging events, but is quasi-stable. The insulator is also challenged by random occurrences like condensation, frost and onshore gales. These add water, ionisable material or both, which, depending on the design of the insulator, either will or will not carry the surface conductivity into a range where flashover can develop.

Remedies are thus needed either because the resident layer is dangerously high in equivalent salt density, causing flashover under the normal low-level challenges which are frequent, or because the challenges are more severe or more frequent than had been assumed. Too high a resident layer can arise either from misjudgment of the local severity or wrong choice of insulator: not uncommonly the severity is altered by the actual installation of the electricity supply system, which triggers industrial development with consequent pollution. Random challenges are difficult to assess because more than one process may be operating. The catastrophe in England[211] in 1962 was caused by a prolonged freezing fog following a long period without rain; highly conductive solution was thereby trapped at the surfaces of the insulators. Neither drought nor fog alone would have caused the flashovers; their combination was clearly highly improbable.

Wide differences thus arise in the type of remedy which is needed, depending on what is causing the flashover and what the relative costs are of outage and remedy. The case of the excessive resident layer calls for a permanent long-term remedy; that of the rare but severe challenge may well be met by a palliative.

The candidate remedies for flashover affecting lines include the installation of more or better insulators, the adoption of V or X sets, the fitting of devices to enhance the creepage path or the use of resistive glaze. Palliatives could include greasing or live washing. For substations the choice is much smaller. Although in principle longer or better insulators could be fitted, in practice these are likely to be ruled out by considerations of engineering and cost. For example, to increase the height of a set of busbars, so as to accommodate longer support insulators, calls for major civil works, the revision of the whole insulation co-ordination and possibly the replacement of very expensive pieces like transformers, in the cost of which that of the insulators is usually a trivial percentage. Reduction of pollution severity by enclosure is also expensive as, of course, is replacement by metalclad apparatus of the whole substation. In general, the choice is between greasing and a fixed live-washing plant, both of which have serious disadvantages. Limited use of resistive glaze, coating with solid hydrophobes and addition of creepage extenders are alternatives which are as yet not fully established.

It is important to realise that not all remedies are mutually compatible. It is undesirable to grease some insulators in a live-washed substation or to subject resistive glazed insulators, on lines, to jet washing.

14.2 Optimised insulator shapes and creepages

Sufficient has been said in previous Chapters about the likely improvements which follow increases in length of insulator, increases in creepage-path length alone, inclination from the vertical and replacement of longrods or line posts by cap-and-pin strings to allow assessment of their relative values as permanent remedies for flashover.

Reduction of surface gradient is by far the best step to take, as a remedy, provided it can be done without serious side effects: it is worth repeating that some 10% decrease in surface gradient is as effective as a halving of surface conductivity, in general.

On overhead lines it is often possible to accommodate an extra 20% of units above those in a vertical string, in legs of a V set, while horizontal strings can usually be extended with little difficulty. Within substations an inverted-V arrangement can sometimes be used for busbar support duty instead of vertical posts, but here the additional creepage may be bought at the cost of increased risk of heavy-wetting flashover (see Section 14.3.1).

14.3 Insulator washing

The principle of washing, either with or without disconnection of voltage, is to reduce the surface resident layer without provoking flashover either of the insulator which is being washed or of its neighbours or, of course, of the jet or spray. In dead washing the risk arises from overspray of polluted water onto adjacent live insulators: this is not negligible when high washing pressures are used.

Live washing uses either jets or sprays, in general, although a Japanese system adopts a water-curtain system broadly copying natural rain[212] (Fig. 14.1). Last and Stalewski[106] have reviewed the history and scope of washing; a selective bibliography is given in Appendix C.

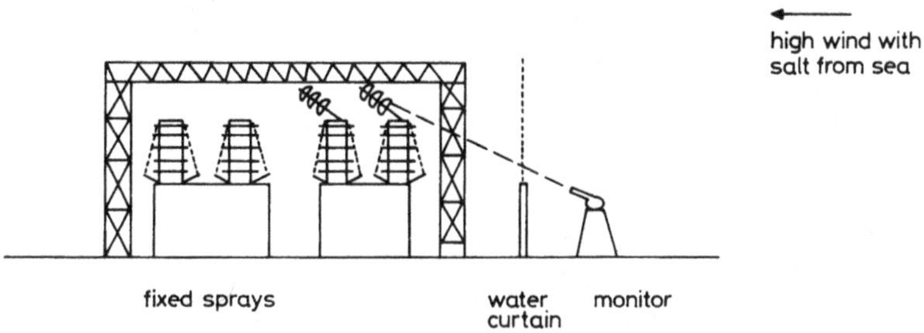

Fig. 14.1 *Japanese live washing: Combination of systems (Reference: NGK TN 70048, June 1970)*
Water curtain intercepts and dilutes wind-borne salt, in typhoon conditions
Sprays and monitors are provided for ground-mounted and aerial insulators

Jet washing is effective in removing solid pollution and is appropriate to adhesive deposits. Spray washing, on the other hand, uses greater quantities of water but is more efficient than jet for removing soluble material. Mobile washing equipment is confined to the jet system in general, although mobile water-tanks are occasionally 'plugged into' permanent spray nozzles.

Typical equipment comprises a 5000-litre tank, pump, hoses, nozzles and access boom, all mounted on a vehicle. Pumping pressures range from 14 to 40 kg/cm^2, with nozzle pressures somewhat lower. Flow rates are of the order of 100 litres/min. To reduce the hazard of current leakage along the jet, common practice is to use an automatic interrupting system which delivers a series of chopped sections of water. A minimum value of water resistivity is usually specified, upwards of 10 000 Ωcm, with the intention of reducing the risk of flashover at the washed insulator.

Clearances from nozzle to live metal range from 6 m at 275 kV to more than 60 m at 765 kV. At such long distances there is evidently disturbance by wind. Fixed nozzle systems, like that used on the Thames Crossing tower (Fig. 14.2) require remote-aiming and position-monitoring equipment. In all cases the string or post is washed from bottom to top, to minimise the risk of cascading by polluted wash water.

Fig. 14.2 *Remote controlled jet washing: Thames Crossing, CEGB, England.* Clusters of insulators on a 190 m-high tower are severely polluted by the plume from a chimney. Servo-positioned nozzles were designed, by A C Stalewski, to live-wash these. (Photograph by courtesy of SE Region, CEGB)

Automatic spray washing systems are used in many substations which are subject to coastal pollution. A typical multi-nozzle system is illustrated in Fig. 1.2. The wash is initiated either on command or, in some cases, by an automatic detector which senses the level of pollution. Very large flows of water, of 2000 litre/min or more, are used, all of it of controlled resistivity and consequent high cost.

Although washing is effective in removing pollution, the total operation, especially of a large substation installation, has many disadvantages as well as high capital and running costs. The fault rate of the washing plant tends to exceed that of the insulators: there will be thousands of nozzles, kilometres of pipework, dozens of valves, all subject to corrosion and statistical malfunction. Pumps and water-processing plant will need regular maintenance and replacement. Less obvious is the change in local microclimate, leading, in some cases, to much enhanced humidity-induced failure rates in other pieces of equipment than the insulators and washing gear.

The most serious defect of large-scale live washing is, however, the provocation of heavy-wetting flashover. Some types of insulator, especially those with closely spaced sheds of about equal diameter and tapered bushings or busbar supports which are inclined, flash over readily when lightly polluted and heavily wetted. Elementary analysis[142] shows that the critical burden of pollution required to produce flashover of an insulator whose creepage path is spanned by a cascade is less than 1% of that required to produce normal surface flashover in light wetting.

The author and his colleagues investigated the multiple flashovers which occurred in spray-washed coastal substations shortly after the commissioning of spray-washing plant, and showed the extreme vulnerability of some shapes and arrangements (Fig. 14.3)[142]. A remedy turned out to be needed for flashovers caused by the washing, itself a remedial measure! Solutions to the problem were found, respectively in the installation of insulators having better cascade resistance and in the Booster shed.

14.3.1 Booster shed

The investigation comprised both pollution and simulation of the conditions likely to arise during live washing. Some of the insulators (Fig. 14.4) were vertically mounted while others were inclined. The simulated conditions included: normal washes under different pressures; short washes of a kind likely to arise from valve malfunction; impulse wetting with pressure falling during the wash – a severe condition known to arise from the discharge of a pressurised pipe; heavy rain with side wind.

A variation of the salt-fog test was used in which the standard fog was used as a pre-pollutant for 15 min, a stabilising time of a further 15 min was allowed, and finally the simulation of the wetting condition was applied. In this revised test, values of WPS or 'withstood pre-applied severity' were obtained, not equal to the withstand salinities in the normal Fog Test but valid as indices of performance under wetting.

Selected results are compared (Fig. 14.5) which show the very weak performance of some insulators, especially those with close-spaced equal sheds, under all washing conditions but especially malfunctions. The vulnerability of other types, particularly inclined tapered bushing shells, to rain is also striking.

The improvements resulting from the fitting of Booster sheds to the different types, in numbers from 5 to 10 for 400 kV lengths, are remarkable (Fig. 14.5): increases in tolerable pre-wetting deposit of salt are typically fourfold, but

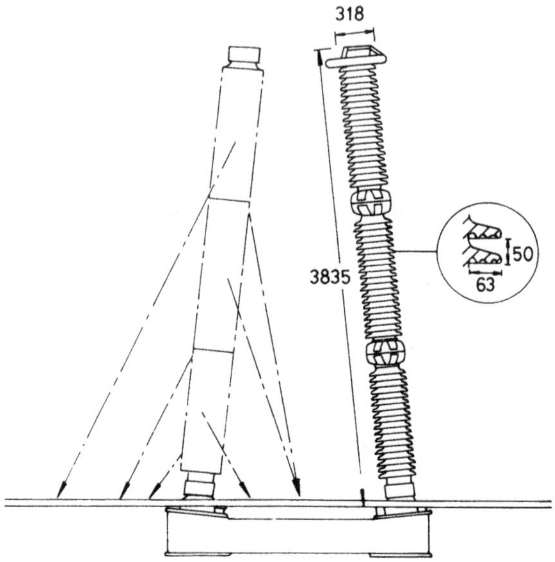

a

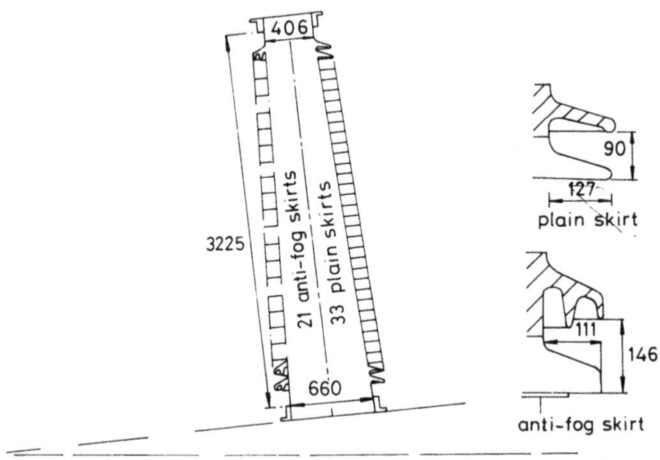

b

Fig. 14.3 *Insulators vulnerable to live washing*
a Solid core inclined post
b Inclined tapered bushing
Dimensions are in mm
Closely spaced, equal diameter sheds (*a*) and taper cancelled by inclination (*b*) increase liability to flashover under heavy wetting

values of up to 100-fold were measured. A demonstration of the efficacy of the devices was made on a vertical insulator (Fig. 14.6(b)) carrying voltage at an intermediate electrode. The lower section was fitted with the devices, the upper section, 150% of the length of the bottom section, was left bare. Under pre-pollution and side wetting, flashover occurred 40 times on the upper low-stressed and lesser-wetted section, but not on the bottom section.

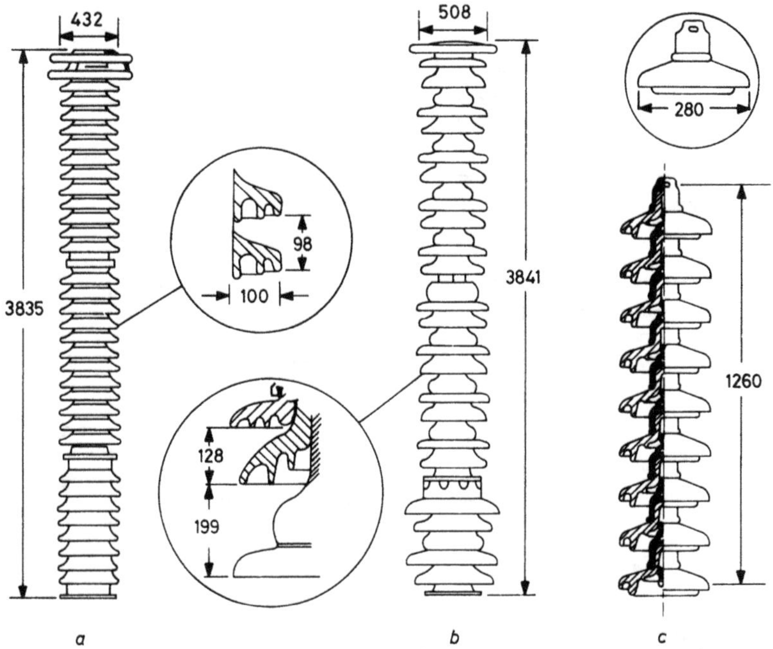

Fig. 14.4 *Insulators tolerant of live washing*
Pedestal posts (*b*) perform well because of their open profile and intermediate metalwork: multiple cones (*a*) are less good. Suspension strings of cap and pin are highly tolerant (*c*).
Dimensions are in mm

Detailed laboratory investigation showed that the action of the Boosters was more than mere water-shedding. Metal parts were ineffective, as were insulated Boosters with multiple punctures between faces. Transparent cover plates revealed that arc-extinction processes were occurring beneath the Boosters. Subsequent tests, using fuse-wire initiation of arcs under Boosters, were made by closing a CEGB 275 kV circuit onto the fitted insulator in an outdoor system-fault test. The high-power fault arc, carrying kiloamperes of current, was ejected from beneath the Boosters without damaging the porcelain substrate.

Details of the optimum number of Boosters, their spacing from the insulator, required overhang and choice of material were established[142]. Their protective value is now recognised from several major installations.

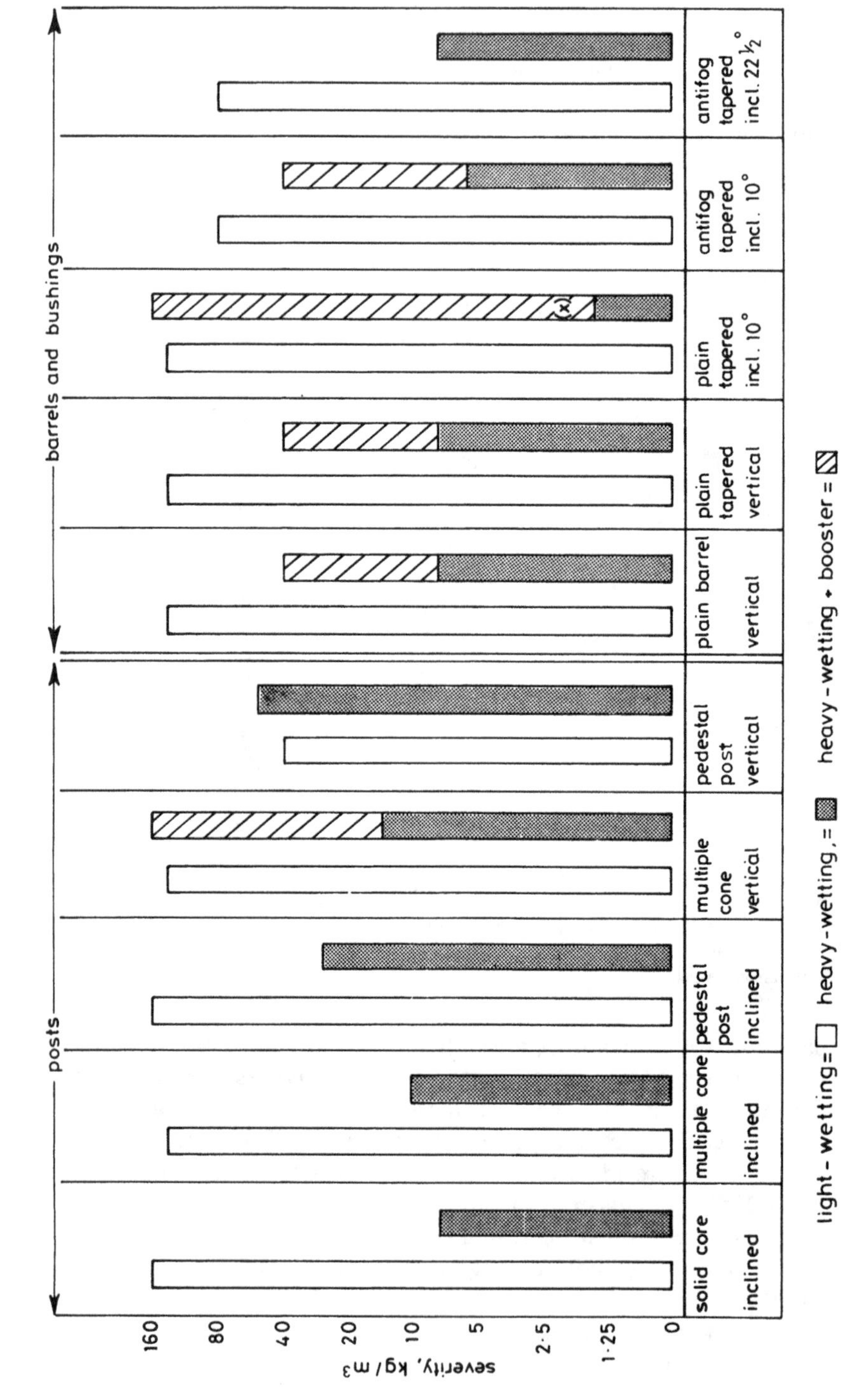

Fig. 14.5 *Tolerable pollution severity under light wetting and worst-case, heavy-wetting/rain*
Extreme degradation of performance is shown by inclined and tapered plain-shed bushing (x)

14.4 Surface treatments

The value of a mobile surface, unwettable and able to absorb pollutants, was recognised in the oil-bath insulator even before the first transmission line was designed (Chapter 1). Many kinds of surface treatment, by oils, greases and pastes, have been applied since, invariably with some degree of success but usually carrying disadvantages, especially damage to the substrate[213].

Mobile coatings act to prevent flashover in two separate ways, respectively to reduce the tendency for water drops to coalesce into a continuous film and to encapsulate particles of solid pollution, thus preventing their going into solution and adding to the surface conductivity. Both the principal classes of coating, silicone pastes and petrolatum gels or hydrocarbon greases, have been extensively used in England and their properties compared[213] (Table 14.1).

Silicone pastes are mixtures of silicone oils with carrier powders, usually silica flour or similar inert minerals with large specific surface area. They are soft and

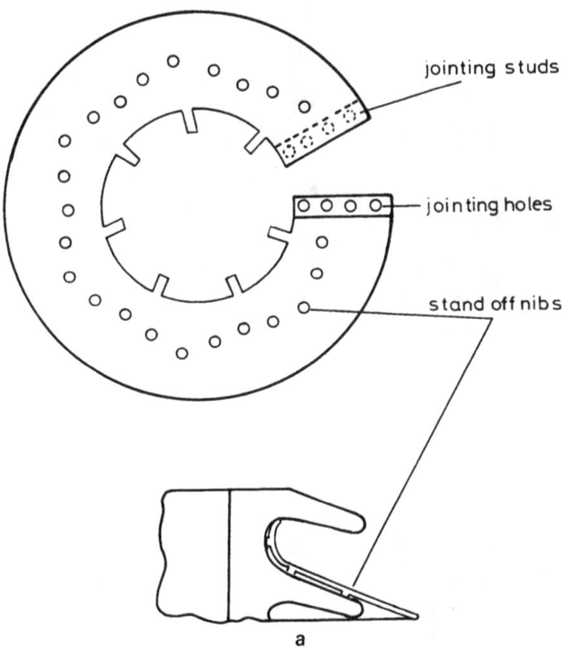

Fig. 14.6a *Booster shed and demonstration of its effectiveness*
a Booster shed: discharge-resistant polymer, fitted as truncated cone and spaced by nibs and tongues

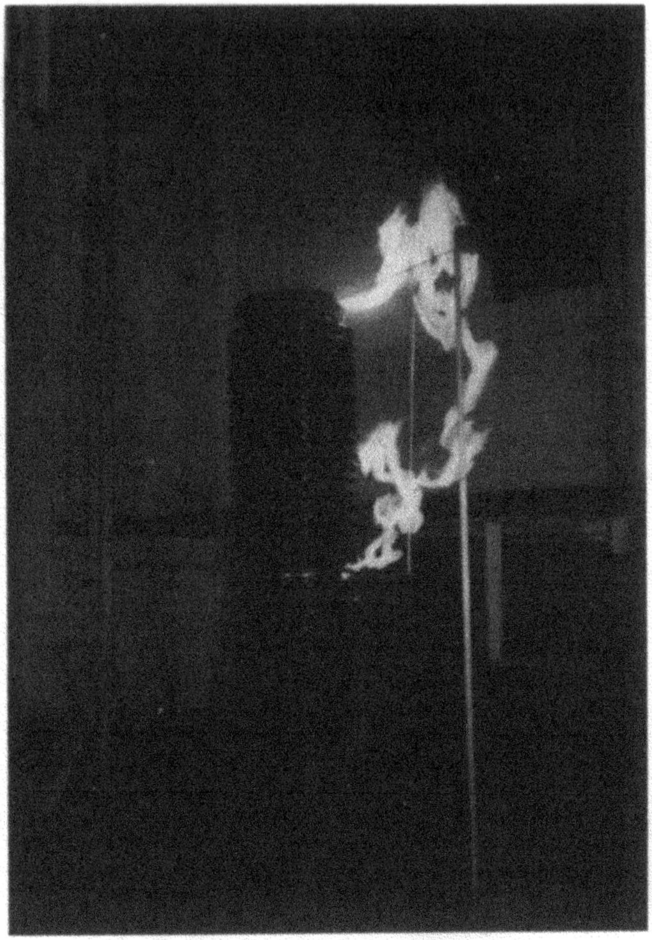

b

Fig. 14.6b Long upper section flashes over under wetting: short lower section, boostered, does not (See p. 201)

easy to apply, and are not subject to sliding in hot climates since their viscosities are about constant from $-50°C$ to $+200°C$. They slowly lose oil, by draining under gravity in normal conditions; under discharges the oil is volatilised or decomposed, leaving debris which can become electrically conductive when wetted.

When in good condition, silicone pastes are highly effective. Particles of solid pollution are rapidly coated with silicone oil: the time depends on the viscosity of the oil, but may be as short as minutes[214]. This isolates them from any condensation or other surface water. Water droplets are not encapsulated but held separate from each other, until the local electric-field intensity is sufficient

Table 14.1 *Mobile coating materials: Properties and performance**

Coating	Composition	Sliding temperature	Performance: severe pollution normal stress
Basic hydrocarbon grease	Oil, wax	48–70°C	Survived flashover > 4 years Poor surface state Slid in heat (3 mm layer thickness)
Controlled-viscosity gel, petrolatum based	Oil, wax, organic additive	115°C	Survived flashover > 4 years Fair surface state Did not slide (3 mm layer thickness)
Silicone paste⊕	Methyl polysiloxane oil, silica flour carrier	Did not slide in 1 mm layers; 3 mm slumped from some surfaces	Flashover in < 3 years Surface showed dry tracks, with damage to substrate, at $1\frac{1}{2}$ years (0·3 mm layer thickness) (1 mm layer survived > 3 years before flashover)

* Test results from CERL Insulator Testing Station, Brighton

⊕ This paste was developed before 1970 and may not be representative of modern materials.

to override the surface-tension forces, causing them to disintegrate, often explosively. As the paste saturates with solid dirt and loses oil, with time, it progressively loses the ability to smother new arrivals. Heavy local sparking then begins to destroy the coating, and sometimes to damage the substrate.

Thicker layers of silicone paste would evidently confer longer effective lives, but these are impracticable, first because of the mechanical instability of layers more than a few millimetres in thickness and secondly because of the high costs of the material (at the time of writing, some 100 times those of petrolatums).

Petrolatums basically contain hydrocarbon waxes and oils, but some also embody polymers or other large molecules, the purpose of which is to reduce the variation of viscosity with temperature. They act differently from the silicones, not smothering arrivals so rapidly because they have less mobile component. When discharges or heavy sparking occur they melt locally, encapsulating the dirt and presenting a fresh surface, which is highly non-wetting.

Because of their cheapness they can be applied in thick layers – 3 mm is recommended[213] – and their tenacity makes them less likely to slump than silicones. However, their principal disadvantage is that all have relatively low sliding temperatures; i.e. the effective viscosity at the grease/substrate interface falls rapidly as the temperature increases. Surface temperatures of porcelain insulators, even in the UK, have been known to exceed 60°C: it is evident that greases which slide in the range 50–65°C, as was the case with the original materials, will be troublesome in temperate climates and useless in tropical ones. New materials now offer sliding temperatures higher than 100°C with, however, certain limitations on modes of application.

The hydrocarbon materials do not suffer from the dangerous deterioration that affects powder-based pastes, but they, too, may cause glaze damage if applied too thinly or allowed to remain after becoming overloaded with dirt. A maximum loading of 20% dirt is recommended[213], but the decision to change may be made earlier because of loss of material by dripping (Figs. 14.7 and 14.8).

The greatest objection to hydrocarbon greases is the cost and unpleasantness of removing spent layers. The mass of grease in a complete substation may amount to tonnes, which usually has to be removed by scraping; the unavoidable residues present serious problems of safety and even of security against fire.

As short-term palliatives, however, greases and silicone pastes are highly effective, increasing the flashover voltage in some circumstances by three to five times[215].

Striking improvements have also been observed in the behaviour of polluted surge arrestors following the application of grease to the outside of the housings. The effect, in this case, is to suppress the heavy discharging across dry bands which occurs with uncoated housings and which, capacitively coupled into the interior, ignites the spark gaps separating the nonlinear resistors and may lead to bursting of the assembly.

Fig. 14.7 *Silica tracks in exhausted silicone paste*
Local leakage current in tracks damages substrate

Special circumstances govern the use of mobile coatings in reducing leakage current over live-working equipment (Chapter 15).

14.5 Use of solid hydrophobes on surfaces

There has been confusion between mobile coatings, which offer self-renewing hydrophobic surfaces, and solid substances like fluorocarbons and silicone elastomers, which, although water repellent and of low adhesion, are subject to both deterioration and damage when polluted and electrically stressed.

Early work by the author's group showed that it was practicable to coat porcelain cap-and-pin discs with layers of PTFE (polytetrafluorethylene), using special priming and sintering methods. The coated insulators behaved well, under natural marine pollution at Brighton, for some one and a half years, during which their counts of leakage current impulses were very much lower

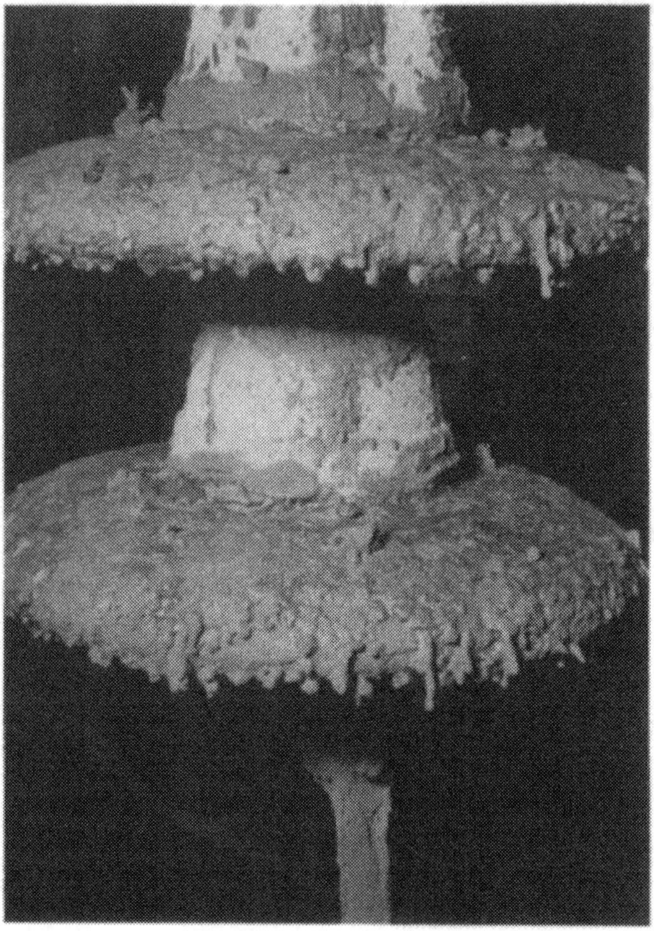

Fig. 14.8 *Hydrocarbon grease after 3-year duty in pollution*
Grease is in bad state: substrate is undamaged

than from normal, uncoated discs of the same shape. It was then noted that the coatings, which were generally not more than 1 mm thick, showed signs of deterioration around the highly stressed pin regions of the insulators. They had evidently been eroded by discharges or related chemical action. The surge counts rose about this time, and the counting rate, after a few months, was significantly worse than from untreated controls.

Porcelain insulators coated with silicone elastomers, of room-temperature vulcanising class, gave similarly disappointing performances at Brighton in severe saline pollution. In bulky industrial contamination at West Thurrock on the Thames Estuary, however, good performances were obtained for up to seven years of exposure from similarly silicone-coated units. Orbeck and his colleagues[216] have reported significant benefits from silicone elastomeric coatings: it is certainly a fact that the leakage-current counts from insulators embodying thick housings of some hydrophobes, especially PTFE, remained consistently low in tests at Brighton, at least in comparison with those from porcelain, glass and many polymeric types having higher surface free energies.

Two facts operate against applied coatings which are non-mobile: some means has to be found to bond them to the substrate, and none can give benefit when coated by a blanket of contaminant. Considering large substation insulators, for which improved performances are most urgently needed, it is seen that the practical objections to the use of fluorocarbons, such as the need for priming and for high sintering temperatures, are unlikely to be overcome. If silicone elastomers are used, the migration of water vapour, to which all silicone rubbers are highly permeable, is a possible source of interfacial weakness. The blanketing effect can give rise to severe erosion when discharges are trapped anaerobically beneath an inert layer, as well as defeating the desired reduction in surface free energy.

Consideration of these facts led the author to devise the Hybrid-insulator principle, by which the benefits of a polymeric surface can be combined with the reliability of porcelain as a mechanical element[13].

14.6 Hybrid insulators

The required properties of a polymeric coating on a porcelain insulator are: sufficient thickness to accept discharge erosion without dangerous penetration, low value of surface free energy on the exterior face, invulnerable water seal to the porcelain. A simple way to obtain these properties is to use a pre-formed part which can be applied to the substrate by shrinking. If such a part is to be made it can as well provide the required long-creepage profile: all that remains for the porcelain to do is to act as mechanical support or housing, for which purpose it can be simply cylindrical with no sheds or skirts of its own.

Important advantages result from a decision to use a very simple ceramic part: these include very large reductions in weight, size and cost, and the achievement of near-theoretical values of mechanical strength by elimination of manipulations of the unfired ceramic body, such as turning, pressing etc., all of which are potential sources of internal flaws.

The remaining problems in evolving a successful Hybrid insulator are how best to terminate the polymeric part and how to ensure invulnerable sealing. Clabburn[13] proposed the use of a refractory band at each end of the polymeric part, by which the polymer would be spaced clear of the terminal metalwork and thus freed from the chemical and electrical stresses, of especial severity at that frontier. Investigations by the author, with Atkins, Rowe and Robles[217], showed that a visco-elastic material, rather than a chemically bonding adhesive, offered the best prospects as interfacial sealant.

Hybrid insulators have now been developed for use up to transmission voltages, and comprehensive tests have been made on the interfacial material[76]; particularly interesting differences have emerged between the behaviours of interfaces where the substrate is ceramic, as in the Hybrid, and where it is fibrous composite, as in the polymeric insulator. The Hybrid interface is remarkably stable, even in the presence of artificial faults which cause the fibrous rod to fail rapidly (Fig. 14.9), at least in the case where the outside of the polymeric part is not heavily polluted. There are grounds for hope that interfacial failures will not afflict Hybrids as they do polymeric insulators.

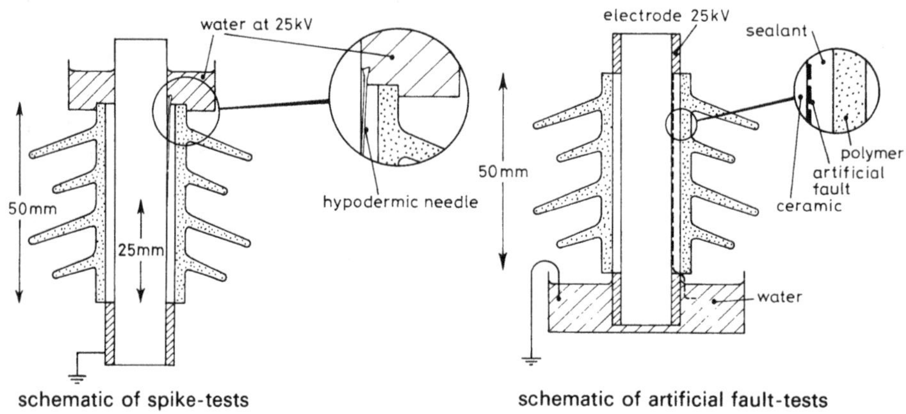

Fig. 14.9 *Spike- and artificial fault-tests*
Fibrous cores fail rapidly: interface with ceramic is relatively stable

The combination of a fairly thick polymeric housing and simple ceramic substrate structure has been found to give incidental advantages, one of which, almost complete resistance to impact damage, has proved valuable as an anti-vandal device (Section 8.2).

Variations of the Hybrid principle have included the addition of ancillary skirts to porcelain insulators[164] and the use of ceramic refractory bands in fibrous-cored composite insulators. Added skirts, or creepage extenders, are fitted to conventional shedded profiles in such a way as to increase the creepage length and give protection against inter-shed sparkover caused by water droplets (see Fig. 17.5). Improvements fully in agreement with calculation are obtained[218]. Ceramic sleeves, fitted immediately adjacent to the metal terminals, have been proved to give good resistance to electrochemical and discharge attacks both on fibrous-cored tension insulators and on optical-fibre conduits[76,164] (Fig. 14.11).

In all these applications the reliability of the sealant is crucial. Recent experiments are encouraging (Fig. 14.9), although some propagation failures are probably inevitable.

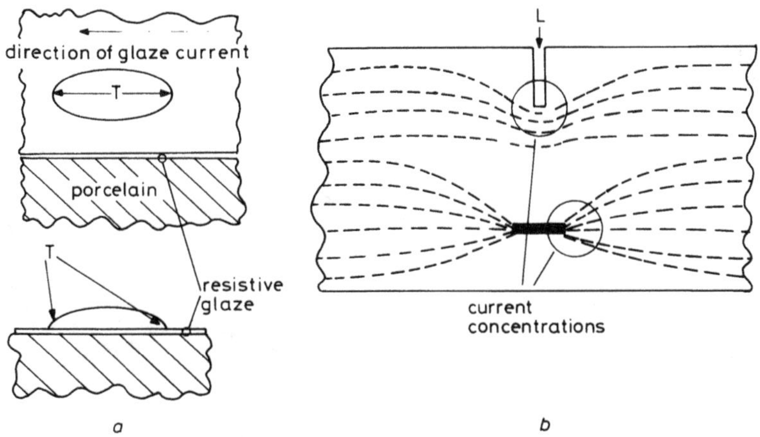

Fig. 14.10 *Damage to resistive glaze*
a Current transfers at high density from glaze to conductive droplet at T
b Open-circuit line L, across glaze current flow, spreads laterally because of current concentration at end. Parallel short circuits also propagate

14.7 Resistive glazes

The demonstration by Forrest in 1936, that nonuniform voltage distribution along an insulator is associated with flashover, led to various attempts at stabilisation. These included the use of parallel resistors connected across units in a string, and following conventional practice for safeguarding stacked capacitors on direct current. One proposed form of resistor was a semiconductive coating or glaze applied to the insulator unit itself[219].

Semiconducting glazes, primarily for reduction of radio interference, had been patented in 1939: these included metals on, or in, glazes, and mixed oxides of iron and manganese, applied to the body and covered with a non-conductive glaze. In England, Vose[220] and Gillam[221] successfully used resistive glazes for improving flashover performance as early as 1946. Very large increases in flashover voltage at a given severity of pollution were obtained[222], at least a doubling in many cases. Were it not for the severe corrosion effects (Section 13.2) which limit the working lifetime of resistive glazes to a few years in severe conditions where they are needed, there is little doubt that they would be almost universally applied today, instead of being confined as a remedy to rare cases of pollution flashover not otherwise curable.

Resistive glazes cannot act, as their inventors apparently believed, by swamping the leakage current. Practical stabilising currents are of the order of 1 mA, being limited by temperature rise and cost of wasted energy. Such currents are negligible in comparison with those of between 100 mA and 1 A which flow as leakage just before flashover. The glazes' function seems to comprise at least three important features, respectively gentle heating to combat condensation of water, maintenance of drying current even when the pollution layer is interrupted, and suppression of the primary discharges across dry bands from which flashover propagates on ordinary insulators.

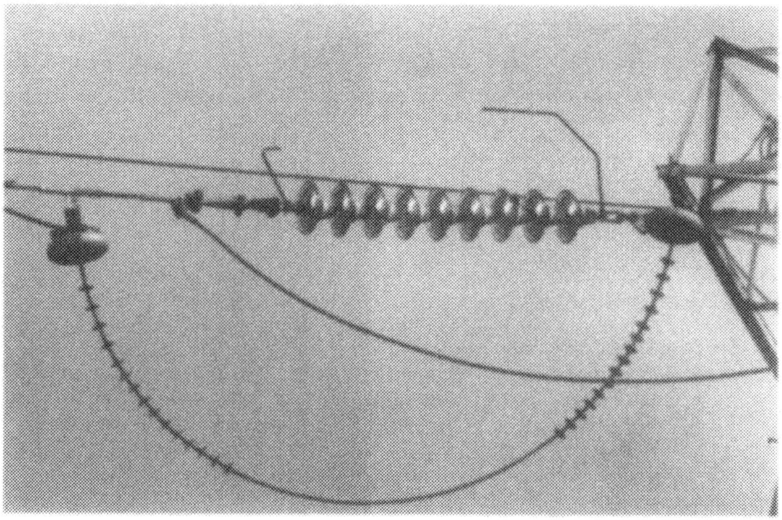

Fig. 14.11 *Optical-fibre conduit embodying ceramic terminal sleeves*

The standing powers which cause heat in the absence of leakage current are small, of the order of 0·5 W/kV system, giving temperature rises no more than a few degrees centigrade. These may have some value, for example in delaying deposition of morning dew, a common cause of flashover in many desert areas, but are irrelevant in heavy fog or drizzle.

The continuous drying current is an obvious difference from that on ordinary insulators. When dry bands form, as they do on both classes of insulator, sinusoidal current continues to flow through the resistive glaze, whereas only intermittent pulses flow, in sparks across insulating dry bands. On the resistive glaze the edges of the dry band continue to retreat until either the whole surface is dried off or the dissipated power just balances the latent heat of the vaporised water. On ordinary insulators the retreat stops once the total dry clearance is equal to the applied voltage divided by the electric strength of air. This clearance is a few tens of centimetres at 400 kV. It is shown, below, that large powers may be dissipated in a resistive dry band when wet contaminant continues to arrive at the surface of the insulator: in fog-chamber tests, thermal fracture of the porcelain resulted from continued fogging of a highly conductive glaze.

Spark suppression by resistive glaze depends very much on the manner in which voltage is applied. 'Cold switch on', the very common case where voltage is abruptly connected to a line which has been left in a wet and contaminated state for some days, gives no time for any resistive heating benefits to appear. Lambeth[175] shows that, in such cases, no discharge will occur on development of the first dry band, unless the conductance of the pollution layer is some 30 times that of the glaze. Where voltage has been applied for a long time, a resistive glazed insulator will have dry regions which are hot and wet ones which are cold; neither will support discharges. Abrupt arrival of water or contaminant in sufficient volume to cover the whole surface may cause transient discharge activity as new dry bands develop. Even here, however, the glaze resistors may well act to absorb stored charge in local capacitances which otherwise would cause sparking on ordinary surfaces.

Because of its apparently abortive outcome, research on resistive glazes will be only briefly reviewed. Fuller accounts are given elsewhere[223].

The principal areas of research and development have been in corrosion and its inhibition, thermal stability of glazes and connection of glazes to terminal metalwork. Much work on process control and glaze formulation has also been done by manufacturers, little of which has been published, and by such bodies as the British Ceramic Research Association[224].

Studies using corrosion cells showed that the early glazes, containing four or more conductive components, suffered badly when charge was exchanged between surface pollution and glaze. Ferrites and, later, partially reduced titania were found to perform better, although the latter was found to re-oxidise to its insulating form under discharge heating. Tin oxide, doped with bismuth, was developed by Binns, and has been the subject of most recent application and testing[225].

The mechanism of conduction in tin-oxide systems is electronic, with some ionic contribution at high temperatures. All exhibit negative coefficients of resistivity with temperature, the best temperatures for half-resistivity being no higher than 350°C. It was originally assumed that such fast falls in resistivity would promote thermal instability, and steps were taken to retard them; however, practical tests indicated that some amelioration of high local stresses was provided by them, a potentially valuable feature especially with certain geometries of insulator. Many tin-oxide glazes were also found to have voltage coefficients of resistivity, which acted additively to their temperature coefficients; some were permanently changed by passage of direct current, while others showed upward drifts of resistivity with time, even when unenergised.

In manufacture, control of resistivity proved difficult. The temperature cycle in firing, the kiln-gas composition during cooling, and even the geometry of the glazed piece, were all found to affect the electrical properties. Viscosity, on which both the thickness of glaze and uniformity depend, especially at edges and discontinuities, had to be controlled by additional fluxes, which again had electrical consequences[226].

Termination by ordinary cement, an ionic conductor, was found to be unsatisfactory since corrosion both of glaze and cement commonly resulted. Carbon-loaded cements were used with success; metal electrodes, applied by spraying, were found to corrode. Amelioration of conditions close to metal terminals was attempted by grading both the thickness and the resistivity of glaze in those areas[227].

Investigations in Canada by Nigol and his colleagues led to proposals for glazes having low metal contents and small particle sizes[228]. Unconventional conical shell shapes were also developed, carrying glaze of much higher surface resistivity than the normal 0·1–10 MΩ/square. None of these steps led to satisfactory lifetimes; some caused early failure, by thermal processes, of the porcelain shells themselves.

It is interesting to compare the deterioration effects which are seen on resistive glazes with those on polymers. In both cases some diffuse attack on the surface must be accepted as inevitable. With resistive glazes the random sites occupied by conductive droplets, and the continually advancing and retreating edges of wet films, cause temporary lines of high current density, which are also concentrations of electrolytic corrosion and sometimes of discharge erosion as well (Fig. 14.10). Artificial tests, using wicks wetted with saline solution as simulated pollution layers, fix both kinds of attack in one region and rapidly lead to visible damage. Since glazes are typically no more than 0·25 mm in thickness and have volume resistivities between 10^6 and 10^{10} Ωcm, visible damage is likely to correspond to major alteration in surface resistivity, especially when occurring in lines normal to the current flow (Fig. 14.10). Processes closely akin to dry-band propagation then occur irreversibly in the glaze, leading to an open circuit. Close to metal terminals there is intense generation of electrochemical products of electrolysis, commonly including ions like OH^- and H^+, which attack either the glassy matrix of the glaze or the metallic oxides within it. There is also high probability of anchored discharges, since arcs terminating on metal are able to run stably for some time whereas those between water electrodes are not: both sparks and arcs destroy resistive glazes instantaneously, in the areas – fortunately small – of root contact.

With polymers the diffuse attack comes from discharge erosion at dry bands; localised erosion again comes from electrochemical and anchored-arc attack. Useful improvement results when the leakage-current collecting length is increased at an electrode by use of larger core diameter: the same is observed with resistive glazes. Since electrolytic effects are proportional to quantity of transferred charge, for both polymers and resistive glazes a wet, lightly polluted site is often more damaging than a drier one of greater severity.

The practical conclusion of the analogy is that finite working life must be accepted for both polymers and resistive glazes as an inevitable fact.

14.8 Calculated powers in resistive-glazed insulators

We consider a 400 kV post, glazed to carry 1 mA current when dry. The standing power is thus 240 W in that condition. When wetted and polluted to the edge of flashover the leakage current will exceed 100 mA, correspond to a pollution resistance of 2·4 MΩ. Assuming a total dry-band length of 20 cm the glaze resistance there will be 4·8 MΩ through which will flow some 33 mA. The power developed in this dry glaze will exceed 5 kW while that in the pollution will be less than 2·5 kW: respective specific powers per unit length become 260 W/cm and only 2·6 W/cm. Evidently, the dry bands will extend unless deposit rates exceed 100 mg/cm (length of insulator).

Chapter 15

Insulators for special applications

15.1 Scope of chapter

Although insulators for duty on outdoor power lines under alternating voltage represent the overwhelming majority, special applications exist where the required properties are different and sometimes difficult to achieve.

Important cases of this kind arise on railways, in electrostatic precipitators, in DC transmission and in live working.

15.2 Railway insulators

Early electric railways operated on DC at voltages up to 1·5 kV or on AC, sometimes at low frequency such as $16\frac{2}{3}$ Hz, up to 15 kV. Recent schemes have had to cope with large loads and high speeds, for which power frequency at 25 kV has become standard. The special circumstances of railway insulators include: high required reliability, since mechanical failure can lead to dropping of the traction conductor; severe and unusual pollution from steam and diesel locomotives as well as by metallic dusts from the brakes and pantographs; in some cases direct contact between insulator and pantograph at speeds of order 200 km/h, coupled with needs for small mass and limited diameter; in general, poor or limited natural rain washing; high vulnerability to impact damage in service and from vandals.

Goldring *et al.*[229] have reviewed the development of insulators for the British Rail system since 1956; recent work, especially on polymeric designs, is well described by Bradwell and Wheeler[101,230,231]. The main classes of railway insulator are top ties and struts, used to support catenaries from posts or gantries, section insulators, used to separate the traction conductor into lengths isolated from each other, and insulators for use in tunnels, where clearances are tight and special mechanical and electrical conditions prevail (Fig. 15.1).

Top ties and struts are commonly porcelain rods, although glass discs and pedestals are also used as well as polymeric types, on a small scale at present.

Shapes are not markedly different from those used on power lines, but some confusion has arisen over required creepage lengths. Since railway systems are single phase, an insulator which carries 25 kV is stressed equally to one on a 44 kV 3-phase power line for which, in heavy pollution, a creepage of at least 25 mm/kV system would be needed, i.e. 1100 mm total. For equal security against flashover, therefore, 1100 mm is needed; whereas much smaller levels have sometimes been chosen, as in Asia, resulting in such widespread flashovers that reductions in traction voltage have had to be applied.

In British Rail, the early designs used creepages as long as 1295 mm, with greasing in addition to combat the effects of steam-locomotive contamination. Recent trends have been to reduce the creepage and improve the self-cleaning by simplification of shed shape[230].

Polymeric top ties for catenaries are based on fibrous cores covered with butyl rubber, ethylene propylene or silicone elastomers, fluorocarbons or, recently, heat-shrinkable mixed elastomers (Fig. 15.1*c*). The fluorocarbon types have been extensively used at lower voltages for many years in Germany and Italy. Because of the deterioration which has occurred with some of these types, both of the housings and by mechanical failure of the cores, test requirements have been imposed, based on work by Bradwell and Wheeler[63], which aim to identify those materials which are vulnerable to 'railway pollution', including the iron oxide produced by brake shoes, and to eliminate internal voids. Defects in the size coating applied to the glass fibres have also been shown to contribute to acid-notching failures, especially when fluorocarbons are used as housings.

One reason for the interest in polymeric insulators has been impact damage, especially from vandal attacks, as already mentioned, on glass and porcelain types. All railway insulators are relatively accessible, since they are always mounted at low level and in some places within easy missile range of footbridges. Porcelain rods or posts may thus need 'buffers'[231]; large-scale attacks on glass insulators in Denmark have led to the successful introduction of Hybrid posts (Fig. 15.1*c*).

Section insulators, over which pantographs run, embody fibrous cores carrying alumina ceramic cylinders of a diameter to match the traction conductor. Fluorocarbon spacing washers are interleaved, and the assembly is impregnated with silicone elastomer (Fig. 15.1*a*). Some designs of section insulator have suffered from acid-notching failure: defective sealing, as well as the size coating previously mentioned, have been identified as causes of failure. The other processes which afflict transmission types also operate[36,37].

Underbridge arm insulators, using fibrous cores covered with either fluorocarbons or silicone elastomers, have been successfully used where air clearances are small. These must be able to flex as the wave in the traction conductor which runs ahead of the pantograph passes them, and to recover without unacceptable oscillation afterwards (Fig. 15.1*b*).

Typical requirements for different types of railway insulator are summarised in Table 15.1.

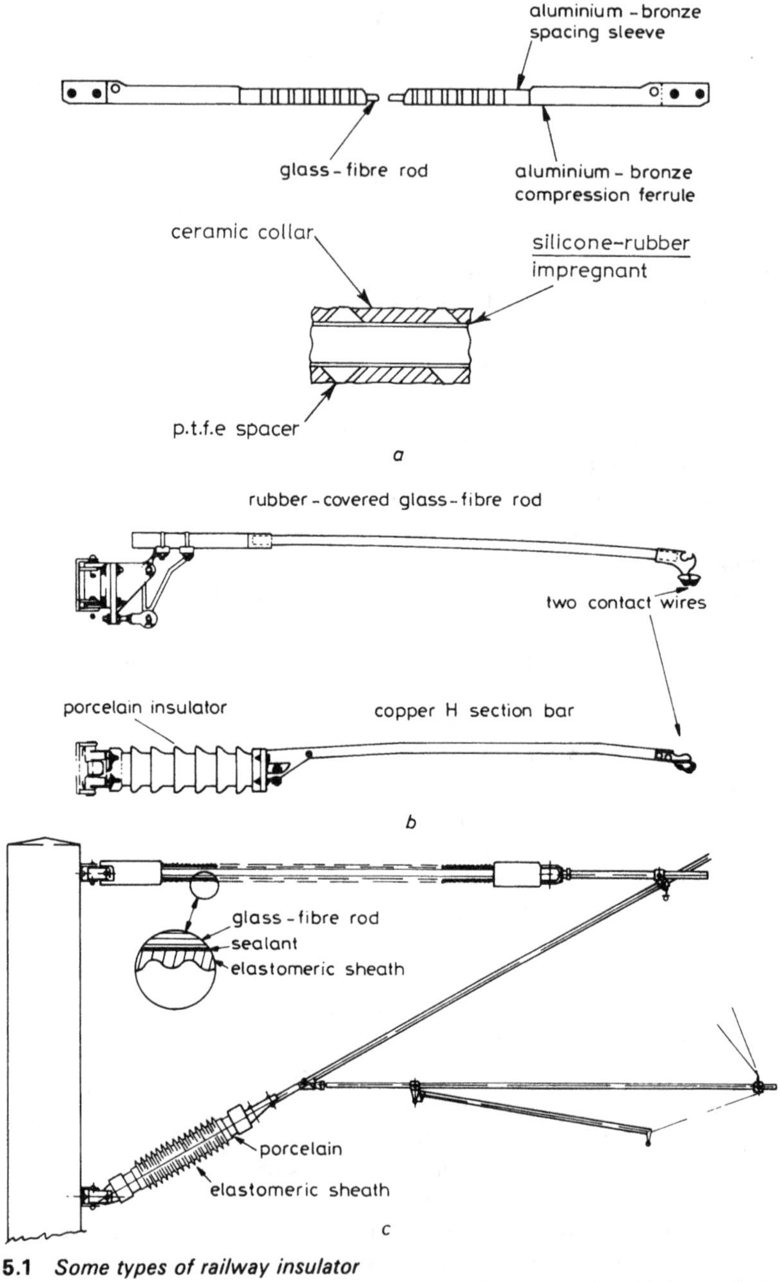

Fig. 15.1 *Some types of railway insulator*
a Section insulator: must tolerate passage of pantograph at high speeds, in direct contact
b Two designs of underbridge arm
c Top tie and strut. Vandal-resistant polymeric tie and hybrid strut are shown. Porcelain, glass and short polymeric insulators are also used.

15.3 Insulators for electrostatic precipitators

Electrostatic precipitators comprise combinations of highly stressed discharge electrodes, usually grids of square rod carrying spikes from which coronas run, with earthed collector plates onto which dust particles carrying ionic charges are deposited. Typical voltages are 50 – 100 kV DC., and pulses are sometimes superimposed. Both collectors and discharge electrodes are rapped by hammers, periodically, to dislodge deposits. The ambient atmosphere is hot flue gas containing water vapour, oxides of carbon and sulphur, nitrogen and some of its oxides, ozone and fly ash, which is mainly siliceous cenospheres of sub-millimetre diameter.

Table 15.1 *Typical requirements for 25 kV railway insulators*

Application	Max. working load	Min. failure load
Top ties, cut-ins	Tensile: 21·2 kN	53 kN
Struts, light duty	Bending: 667 Nm	1·7 kNm
Struts, heavy duty	Bending: 941 Nm	2·4 kNm

Creepages are about 790 mm for normal and 1100 mm for heavy levels of pollution.

Insulators are required to support the discharge electrodes, to locate the conductors which supply them and to carry the voltage through the wall of the enclosed chamber in which the flue gas flows. Since ambient temperatures locally may exceed 110°C, at which porcelain loses mechanical strength and at which significant ionic conductivity begins within it, materials such as fused silica, alumina or cordierite (a ceramic based on Mg, Al, Si), all of which are expensive and relatively difficult to process, have to be used instead.

These materials also offer low coefficients of expansion, reducing the risk of thermal fracture. All require very high temperatures for processing, however – about 1700°C for SiO_2, 1500°C for high-alumina porcelain and 1400°C for cordierite – and the choice of shape is limited.

A recent move for all classes of precipitator insulation, including the wall bushings, is to adopt the Hybrid principle, using single-oxide rods or tubes for the mechanical duty and an applied polymeric sheath for the required profile. The simple cylinders of ceramic, which are all that are required, are relatively cheap and strong: special attention both to the polymeric sheath and the sealant are needed, however, because of the aggressive conditions (Fig. 15.2).

15.4 Insulators for direct voltages

15.4.1 Basic differences from AC condition

An insulator which is DC energised differs from the corresponding AC case: in the mechanisms of pollution and propagation of discharges to flashover, in the processes of surface erosion and corrosion of cement and metalwork and in the aging of the dielectric, especially when this is a highly stressed ceramic. The volume of service experience is much less for DC than for AC insulators and,

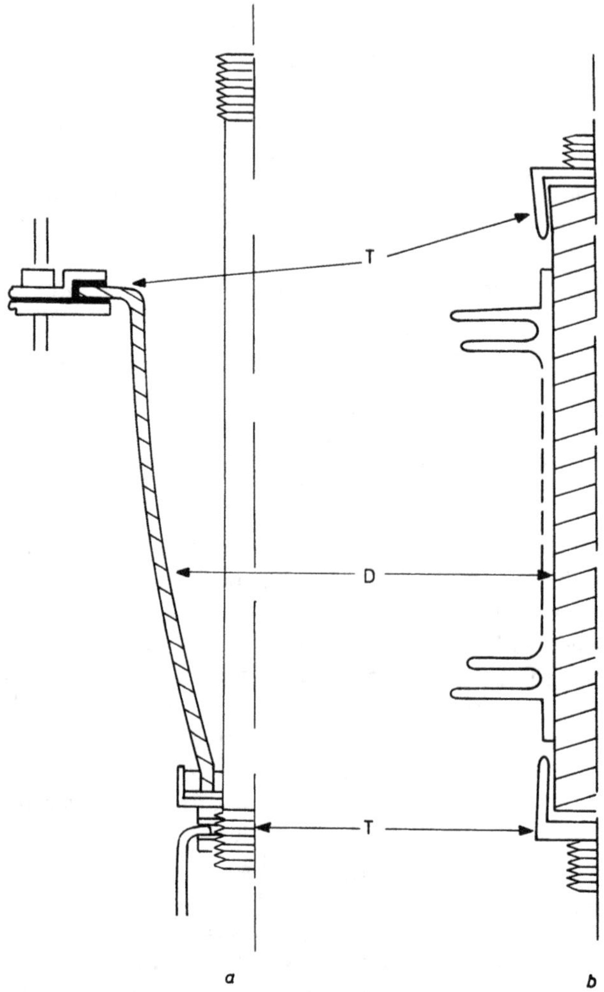

Fig. 15.2 *Precipitator insulators*
a Bushing
b Support
Dielectric D must be refractory material (silica, high alumina, cordierite), which limits feasible complexity of profile
Terminations T must tolerate large expansion differences, cannot embody silicate cements

as mentioned in Section 11.2.3., both the number of DC test facilities for insulators and the reliability of test results have been limited by the difficulty and high cost of providing sources of voltage of realistically low impedance. Insofar as comparisons of electrical withstand voltages under DC and AC are concerned, therefore, the data must be treated as somewhat speculative.

Some differences may, however, be deduced from first principles. Contamination processes must have worse effects on DC than on AC because electrically charged airborne material will drift towards one or other electrode under the steady electric field, depending on polarity. Any discharges on, or adjacent to, the insulator itself will generate copious flows of ions, which, also, will attach to particles and cause their precipitation.

Propagation processes have already been discussed. The absence of voltage zeros on DC, and of polarity reversals, must favour development of primary discharges to complete flashovers. Since the voltage distribution either between units in a string of discs, or along the body of dielectric on a post, rod or housing, cannot depend on capacitance, when DC is applied, but solely on resistance, wider variations must be expected than in the AC case, and hence a greater frequency of primary discharges.

Surface erosion must be worse with DC because of the sustained nature of any surface discharges (which, in fact, are audibly different from AC, as well as more likely to be anchored at one place). Polymeric housings and glass insulators will be expected to suffer especially badly in this condition, not only from direct attack by discharges but also, near electrodes, from the accumulation of electrochemical products. Electrolysis of common salt, for example, produces caustic soda, which rapidly attacks many glasses and polymers[232,233].

For corrosion of metal terminals, swelling of buried metal in cemented fittings and growth of the cement itself, DC is known to be far worse than AC: a rule of thumb is a factor of at least 100, for comparable charge transfer. Polarity effects must also arise: a buried pin, in a disc insulator, which has positive polarity is the reverse of the normal cathodic-protection arrangement. Bursting of heads of cap-and-pin discs under positive polarity has been observed in months, in certain conditions of saline pollution[175].

Changes in ceramic dielectrics themselves under DC stress have been extensively investigated[14]. Since both glass and porcelain contain mobile ions it must be the case that migration will occur under steady stress[234]. Such phenomena as[24] depletion of one species from the area of its own polarity, discharge of ions at electrodes, with physical and chemical consequences, and even instabilities like formation of dendrites, will arise in ceramics as in any other solution, solid or liquid. The practical question is what the rates of such processes will be, at temperatures relevant to normal insulator duty.

In the following Sections the design or selection of insulators for use on DC is considered, respectively under relative flashover liability and relative deterioration rate, in comparison with those for AC duty.

15.4.2 Relative flashover liabilities, DC and AC

As stated in Section 11.2.3., when flashover voltages are compared, using a given type of insulator and severity of pollution, the ratio

$$F = \text{(peak AC flashover voltage)/(DC flashover voltage)}$$

is found to vary with shape of insulator. Salt-fog tests gave values of F from about 1·2 to 1·7[175].

Investigations in Japan[235] using an alternate deposit of kaolin and salt, a test method developed at ASEA in Sweden by Annestrand and Schei[236], showed that F varied with severity as well as shape of insulator, rising above 2·0 for high severities. They showed a polarity effect also, with negative some 10–20% lower than positive.

Large differences in deposit density were found in artificial tests, with local values on DC ten times the average with AC. Studies of effects of shape suggested that distances between tips of skirts should be maximised while retaining long creepage (Fig. 15.3).

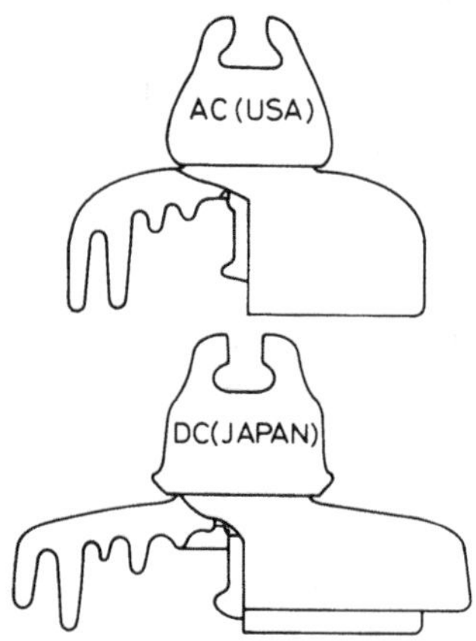

Fig. 15.3 *Different profiles: AC and DC*
Large diameter gives maximum creepage for DC
Skirt tips spaced for DC to reduce bridging by persistent arcs (Data from NGK Ltd.)

New test facilities have now been developed by this group in which the regulation of the DC source has been improved by incorporation of feedback based on thyristor control. Voltages up to 750 kV are available: preliminary results bearing on the behaviour of large bushings have been obtained[237].

Work in France by Pargamin, Huc and Tartier appears to show large differences in behaviour, not only with shape but with the material of the insulator[238]. The results are open to question in that short strings of only three units were used, and that identical shapes in different materials were not compared, both parameters being varied between samples. Different orders of merit were obtained in three test procedures, respectively the IEC 507 standards and the ASEA dust test: furthermore, severities were different in all three procedures. However, in spite of these caveats, interesting results were obtained.

The polymeric insulators generally performed better than the ceramic. The highly convoluted anti-fog shapes and the smooth aerofoil were about equal in tests other than salt fog; as stated previously, the salt-fog test is known to undervalue aerofoil shapes for AC, from which it may be assumed that at least equality of performance with the anti-fog will be obtained in real service on DC also. Pargamin's salt-fog data can be related to other work on the anti-fog shape which gave the best performance[147]: values of F between 2·0 and 2·8 are obtained at a severity of 28 kg/m^3.

An investigation both of relative flashover liability and of insulator-diameter effects under DC was made by Verma[144] for disc types, longrods and housings. He found values of F for anti-fog discs between 1·7 and 2·5 with increasing severity, against 1·4 to 2·1 for both longrods and standard discs. The decline of flashover voltage with increasing core diameter was fast: in the case of the large housings the withstand salinity was only 7·0 kg/m^3, against 40·0 kg/m^3 for a longrod of equal length and creepage at the same voltage.

According to these results, a change in profile from equal overhanging to alternate large/small had little effect: however, comparisons were reported only from horizontal-mounted insulators in which the draining rate of the core would dominate performance.

Table 15.2 *Required specific creepages for DC lines*

Required creepage for AC (mm/kV(RMS), system)	Indicated severity		Factor F		Deduced DC creepage (mm/kV)	
	Type	mg/cm^2	Antifog	Standard/ longrod	Antifog	Standard/ longrod
16·0	light	0·03–0·06	1·8	1·4	35·3	27·4
20·0	medium	0·1–0·2	2·1	1·7	51·4	41·6
25·0	heavy	0·3–0·6	2·4	2·0	73·5	61·2
32·0	extreme	> 0·6	> 2·4	> 2.0	> 94·1	> 78·4

Example

A line operating at 150 kV AC has a satisfactory flashover rate when insulated with standard discs to give 3·0 m creepage. What creepage would be needed for a 250 kV DC line in the same area?

The AC creepage corresponds to 20 mm/kV(RMS), system.

This is $(20 \times \sqrt{3}) = 34{\cdot}6$ mm/kV(RMS), line-to-ground, or $(34{\cdot}6/\sqrt{2}) = 24{\cdot}5$ mm/kV(peak), line-to-ground.

At this severity, the factor F for standard discs is 1·7.

Therefore the required creepage for DC is $(24{\cdot}5 \times 1{\cdot}7) = 41{\cdot}6$ mm/kV or 10·4 m.

Because of the confusion which easily arises between phase voltage, line voltage, peak voltage and DC voltage, when insulation is being selected for a DC line on the basis of local experience on AC, a worked example may be useful (see above). General conclusions on liability to flashover under DC are drawn as follows:

(a) All types of insulator are much more liable to flashover under DC than AC, in comparable conditions. Very long insulators must be provided for severely

polluted places in the case of overhead lines: for large-diameter substation insulators such poor performances are likely that special measures must be considered (e.g. greasing, addition of polymeric parts and washing).
(b) Linearity is obeyed under DC up to about 300 kV, but for higher voltages some departure must be expected, with further increase in liability to flashover.
(c) Some benefit may be expected from the use of polymeric insulators or polymeric coatings, although the effects of weathering and aging are likely to be deleterious.
(d) Special shapes for DC may give modest increases in flashover voltage, at least for overhead-line insulators. For substation types the outlook for special shapes is seen as unfavourable, because of the dominant effect of large core diameters[144].

15.4.3 Relative deterioration rates

Substantial data are available at present only on the decay and fracture of overhead-line insulators. Work within CIGRE by Peixoto, Pargamin, Marrone and Carrara[14] summarises the statistics and reviews the physics of failure under DC.

Average failure rates per annum more than ten times higher than the norm for AC lines are observed for glass discs on lines in Africa, Scandinavia and Italy. These range from 11 to 60 per 10^4 and show trends, when related to time in service, both of a 'weeding out' of defective units and of general aging (Fig. 15.4).

Distinct causes of failure on DC are identified, including fracture by the swelling-anode process, heavy erosion close to the higher stressed electrode and ionic migration within the dielectric. Of these, only the second is a known cause of failure on AC. Remedies could include reduction in ionic migration by increase in volume resistivity of the dielectric, elimination of voids and inclusions from glass (porcelain is inherently granular) and amelioration of corrosion.

Large increases in volume resistivity appear to call for major changes in glass composition. These are not tolerable since they would interfere with the toughening process. Evidently, insulators on DC, in regions where surface temperatures approach 100°C, must suffer large increases in ionic mobility within their ceramic dielectrics. Amelioration of corrosion by the use of sacrificial zinc collars[239], and improvements in protective coating for the metal fittings, are already embodied in some designs. The uncontrollable flow of unidirectional leakage current must, however, override any local galvanic protection, while the protective coating will, in any case, be punctured electrically unless made semiconductive.

It remains to be established whether glass, generally with higher stressed dielectric and consequent higher transfer of ions within it, but always with external surfaces more liable to erosion, will last longer on DC than porcelain, in which the stress is less but the number of mechanical crack-initiating inclusions much greater.

Apparently longrods would be advantageous for use on DC, since they are intrinsically little subject to damage from ionic migration within their dielectrics and contain no buried metal which may corrode. Their generally inferior

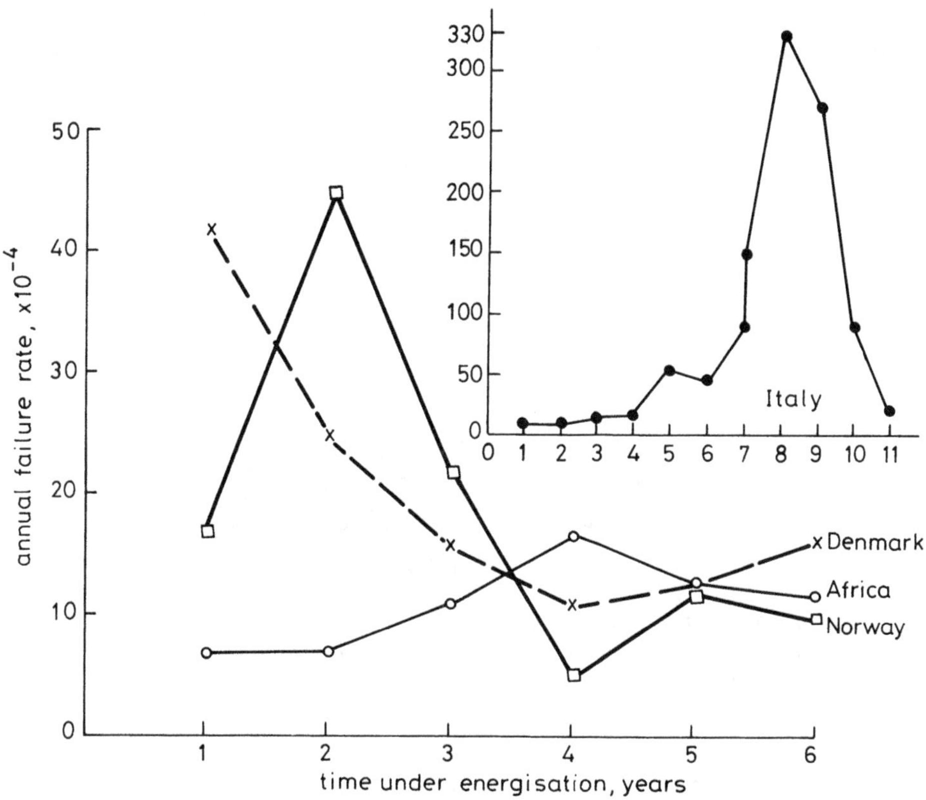

Fig. 15.4 *Annual failure rates: Glass discs under DC*
The starting dates differ: all European lines showed large maxima in years 1977 and 1978

performance in pollution is therefore unfortunate. Although good results have been obtained from longrods on DC in moderate industrial pollution[240], high liability to flashover must be accepted in severely contaminated places. Hirsch[241] has obtained very good results under DC from some types of polymeric longrod: it may well be that these, and perhaps Hybrid longrods, will find useful application on DC lines, provided that the deleterious effects of the pollutant, especially when this is salt, can be overcome[242].

15.5 Insulators for live working

15.5.1 Leakage-current limitation

Live working embraces the whole field of operations which are performed on conductors at high voltage without the normal disconnection and earthing precautions[195]. Both hand-held tools and supports of various kinds are used. The latter may serve to place men on conductors, for example, or to provide temporary support for conductors while an insulator is replaced (Fig. 15.5).

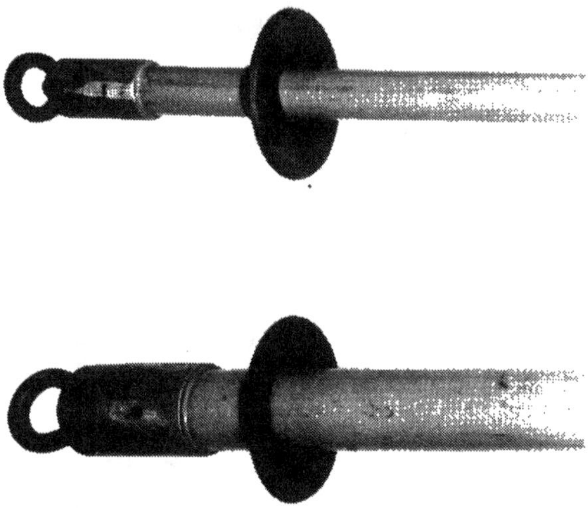

Fig. 15.5 *RBGF tubes for live working*
63 mm diameter used for supports; 38 mm for hand-held tools
Skirts are of retractable modified silicone elastomer

In the case of the tools, called 'hot sticks', the man remains at earth potential and all leakage current flowing along the stick passes through his body. The limit of perception with AC for men is about 1 mA (women, in general, have a different threshold level), and the practice is to keep leakage well below perception in all circumstances. There is little margin for errors: at 10 mA, the 'let go' level, muscle action is inhibited, and at 100 mA, depending on the current path, death may result. Both high-quality material, and restrictions on levels of wetting in which live working is permissible, are therefore essential.

Insulating supports include booms for transporting buckets towards conductors from which men may work 'bare hand' at high voltage, ladders, insulating ties and struts, polymeric chains and ropes. Similar restrictions on leakage apply to these, but with additional requirements for maintenance of mechanical factors of safety: leakage current may cause loss of strength in polymeric structures, and especially in ropes, which may be critically weakened by sub-milliampere currents flowing for only minutes[195].

Flashover of tools or supports is, of course, totally unacceptable: in the case of tools, fatality is nearly certain, while flashover of a support is likely to cause grave injuries to the linesmen who are using it, either directly or by dropping of the conductor. Since high flashover voltage and low leakage level have to be combined in pieces of apparatus having only modest bulk and weight, in the interest of mobility, the problems in design and testing are severe.

15.5.2 Hand-held tools — 'Hot sticks' — struts and ties

The basic insulating material is, almost universally, tubular RBGF comprising a foam core, onto which layers of pultruded axial glass fibres and circumferential filament-wound glass fibres are laid, in a matrix of epoxy resin. The core is supplied skimmed, to allow excellent bonding with the resin, which fills the opened pores, and the tube is extruded and cured in continuous lengths. Tools are made of 38 mm-diameter tube, struts and ties of 63 mm: sizes up to 460 mm are made for booms.

Although the strong epoxy wall is 'diluted' by the foam core, which has low tensile or compressive strength, tensile-failure loads of 66 kN are obtained with 38 mm tubes. Compressive strengths are low because of the ease with which long, thin structures fail by bowing: a load of 8 kN is sufficient to fail a 63 mm tube of length 3 m.

Electrical strengths in the dry state are quoted on the basis of ability to resist 100 kV/ft at 60 Hz with negligible heating. Such stresses, 333 V/mm if uniform, are well below the breakdown value even for air, and are sufficient to detect only seriously defective material, containing water or uncured resin. For satisfying the safety criteria of maximum leakage and maintenance of strength, much more stringent tests are needed and are now in process of specification.

Simple immersion of the RBGF material in water for periods of weeks gave no detectable increase in leakage: tests in fogs of low salinity, however, simulating the conditions to be expected in the field under poor weather or industrial air pollution, showed that alarming currents could be produced at the relevant electric stresses, up to 84 V/mm (Table 15.3). Special wax polishes were developed to reduce these currents, the effects of which, however, were found to be overridden by electric stress, on tubes intended for duty up to 400 kV (Fig. 15.6).

Currents sufficient to damage the surfaces flowed on the larger-diameter samples, some of which flashed over. Surface abrasion and weathering were found, by tests on used equipment, also to increase the values of leakage current in the standard fog.

It was concluded that additional creepage length, in the form of shrunk-on polymeric skirts, would be advisable for some of the larger tubes, and that the booms would require both external shedding and internal reinforcement to inhibit progressive failure of the foam[195]. Both these remedies proved highly successful, showing reductions in leakage and increases in flashover voltage to acceptable levels.

Table 15.3 *Leakage over live-working material in town-water fog (conductivity = 250 μS/cm)*

Sample	Length (m)	Stress (V/mm)	Leakage (peak) (mA)	Remarks
Hand-held hot stick, 38 mm dia., used	2·57	62·3	17·5	currents four times higher than on new sticks
	4·15	38·5	10·5	
Gripper tool, with parallel rod	3·40	47·1	57·7	current sufficient to cause damage
Strut, soaked for 1 week, 63 mm dia.	2·86	55·9	30·0	flashed over: current above 'fatality level'
	2·86	83·9	125·0	
Strut, wax polished	2·86	29·7	1·4	
	2·86	55·9	70·0	
	2·86	83·9	100·0	flashed over
American boom, 460 mm dia.	2·90	55·2	105·0	6.5 mA internal: severe tracking and burning in minutes
'Welsh hat' boom, stacked conical units	2·90	82·7	0·3	no detectable damage
Unprotected square epoxy/glass rod	0·90	54·0	30·0	edges burned in minutes: flashover

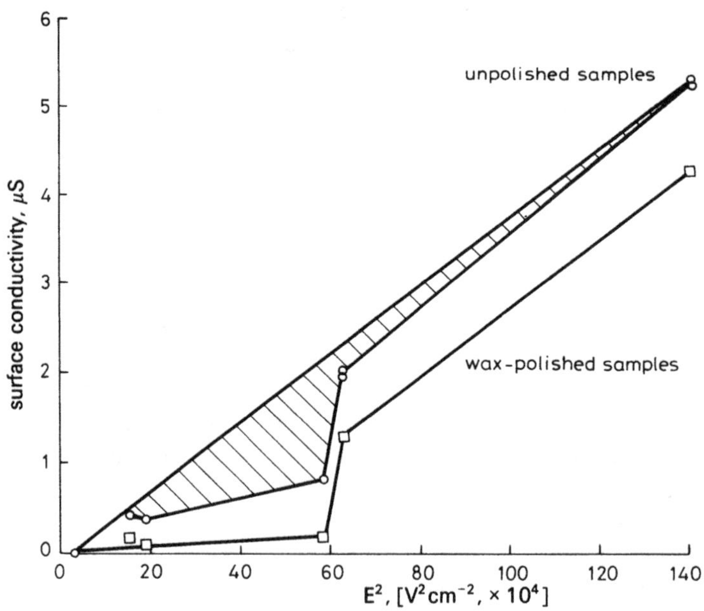

Fig. 15.6 *Surface tension overridden by electric intensity*
Hydrophobic polishes limit leakage current only up to critical electric stress

Disturbing evidence has recently emerged that ingress of water into the terminal regions of live-working tubes may lead to internal flashover. Experiments by the author and his colleagues have demonstrated that damaging propagation of tracks may occur at the foam/RBGF interface if this is highly stressed electrically (as it is in service, both near the live terminal and elsewhere, if the surface is even lightly contaminated). Either use of a sensitive discharge detector to locate incipient tracks, or substitution of solid rods for foam-filled tubes appears advisable.

15.5.3 Tensile supports: Ropes, chains, monofilaments

Although ropes are useful as laboratory insulators in high-voltage experimentation indoors, they are hazardous outdoors. Allin, Butler and Hampton at the Central Electricity Research Laboratories showed 20 years ago that typical water uptakes by weight ranged from 7% to 75%, and that sample lengths fell in resistance by between three and seven decades! Specially treated ropes with water-blocking impregnants have since been developed, but they should be treated with reserve because of their complex dynamic behaviour under load, with strand moving relative to strand, which renders them vulnerable to invasion by water. Externally sheathed ropes are now available which are free of the defects of earlier designs, based on parallel filaments and encased in extruded polymer. These failed rapidly by multiple puncturing and ingress of moisture. The new ropes are water-blocked internally and have convoluted polymeric sheaths of highly discharge-resistant elastomer.

Polymeric chains, moulded without joints in the links (Fig. 15.7), give remarkable performances as insulators. Artificial tests have shown that they offer high withstand salinity, for unshedded insulators, – 5 kg/m^3 at an applied electric stress of 65·6 V/mm – and long life, in natural pollution of high severity, while stressed as if they were interphase ties at 400 kV. Only some classes of polyolefine have given long lives: ordinary nylon, although mechanically much stronger, takes up water and fails rapidly under electric stress.

Currently available mechanical strengths are far too low for any but light duties, failure typically occurring at a few hundred kilogrammes and creep being severe even at smaller loads, but the potentialities are interesting as also is the process responsible for the good flashover performance. This appears to be uniform stress grading by the existence of multiple voltage concentrations at each link-to-link contact: Baker and Voas[243] are currently exploring a similar principle for improving the performance of polymeric longrod insulators by means of multiple 'dry collars' (Fig. 17.5, d).

Similarly interesting and surprising results were obtained in tests in monofilaments. A specially treated nylon filament of 4·7 mm circumference carried a barely detectable leakage current, well below 1 mA, in artificial fog of low salinity. Insufficient current flowed to cause either damage or flashover development at the test stress level of 65·0 V/mm.

The general conclusions on insulators for live working are as follows:

(a) Surface leakage currents are unacceptably high, especially on supports of large diameter (63 mm and above) in wetting or pollution levels likely to arise near the sea or industrial installations, under electric stresses relevant to live work at transmission voltages. Remedies like polishing or added skirts are needed: there can be no question of work in even light rain with ordinary equipment.

Fig. 15.7 *Polymeric chains as insulators*
The linesman is safely supported by polypropylene chains for placement on live 400 kV conductor

(b) The risk of internal flashover following invasion by water is such that monitoring or change of practice is needed.
(c) Multistrand ropes, even when sheathed, are to be used with caution. Polymeric chains and monofilaments are intrinsically safe.

Chapter 16

Interference and noise generated by insulators

16.1 Generating processes

Interference with radio and television (RI and TVI) may arise when electrical discharges run on insulators and inject high-frequency currents into associated conductors, which radiate electromagnetic waves. Audible noise (AN) is generated either by electrical discharges or by an entirely different process, resonance in cavities of insulators, excited aerodynamically.

The types of discharge which generate interference are: microsparks between water drops or metal fittings, the latter especially in cases of corrosion; discharges across dry bands on leaky surfaces; surface corona discharges around highly stressed electrodes such as pins in disc insulators.

The frequency band in which RI or TVI is generated depends both on the nature of the discharge and the electrical parameters of the associated circuit. Microsparks are brief events containing very fast changes in current. In consequence, they are able to generate frequencies from HF up to the microwave range (Fig. 16.1). When occurring between water drops or from drop to metal they are transitory and fairly ineffective as generators, but between metal parts, and particularly when running in or over crystals caused by corrosion, they may give rise to continuous and high levels of interference[244]. Discharges between conducting patches or across dry bands take the forms of sparks and arcs. The sparks 'ring' the local circuits and may generate frequencies up to VHF; the arcs are slow-varying phenomena which give rise to acoustic noise, but are inefficient, for their large relative powers, as generators above MF, being blocked by high resistances. Surface corona discharges are again relatively slow phenomena, incapable of heavy generation at VHF, but principal sources at lower frequencies.

The emphasis has shifted in recent years away from the LF and MF bands, which were formerly dominant and for which much research on abatement has been conducted.

The reason is that the higher-frequency bands have generally supplanted those of MF and below, and that frequency modulation, which is largely immune to interfering signals generated by discharges, is now widely used for radio communication.

Abatement of the generation of interference by insulators thus consists in inhibiting the discharges which are responsible for it. Prevention of radiation of the interfering signals is a matter of reducing, as far as possible, the coupling between the resonant circuit which is excited by the particular discharge and the antenna from which the electromagnetic wave will propagate.

Microsparks have been successfully checked by the use of surface greasing to deal with droplet discharges and also to suppress dry-band discharges on dielectrics, and by bonding of metallic parts or use of conductive greases. The contact pressures are high between caps and pins on normally loaded strings of discs, effectively welding the fittings into solidly conducting entities. On lightly loaded strings such as jumper supports, however, both intermittent contact and invasion by corrosion may occur: such strings are therefore likely interference sources.

Surface discharges may be prevented by hydrophobic treatment, but a highly effective remedy is resistive glazing. This not only inhibits dry-band formation but also gives good unit-to-unit voltage grading, thus removing the overvoltages which cause other types of discharge. It might appear that the installation of a few resistive discs at the live end of an ordinary string would smother capacitive overvoltages, while avoiding the usual power loss which follows installation of a completely resistive string. Dembinski[245] showed, however, that currently available glazes are about ten times too conductive for this purpose.

Both the above-mentioned types of discharge are, to some extent, weather dependent. Water may cause droplet discharges while suppressing contact discharges by virtue of its high conductivity and permittivity; dry bands do not occur in fair weather. The surface corona in insulator-pin regions is, however, a fair-weather phenomenon, the existence of which depends on the design of the insulator and the geometry of the string, inclusive of fittings. Surface corona, if occurring at normal line voltage, would thus generate interference continuously in fair weather, for which reason it has long been the practice to specify higher than normal voltages for corona inception on insulator strings.

Some typical values (Table 16.1) are so much higher than the normally applied voltages that they pose problems for the designer of the insulators, and add to his costs. The stated reasons for these levels have been to give a safety margin, covering such things as minor production faults at normal working voltage, and to cope with the effects of pollution. The first is acceptable, but the second is not. In fact, the mechanisms of RI and TVI generation are so different in the three common states, respectively clean and dry, wet, and contaminated, that measures aimed at improving performance in one condition are liable to worsen it in others[246].

In Great Britain, where the standard 400 kV conductors are run at such low surface gradients that corona-generated RI in fair weather is negligible, it has

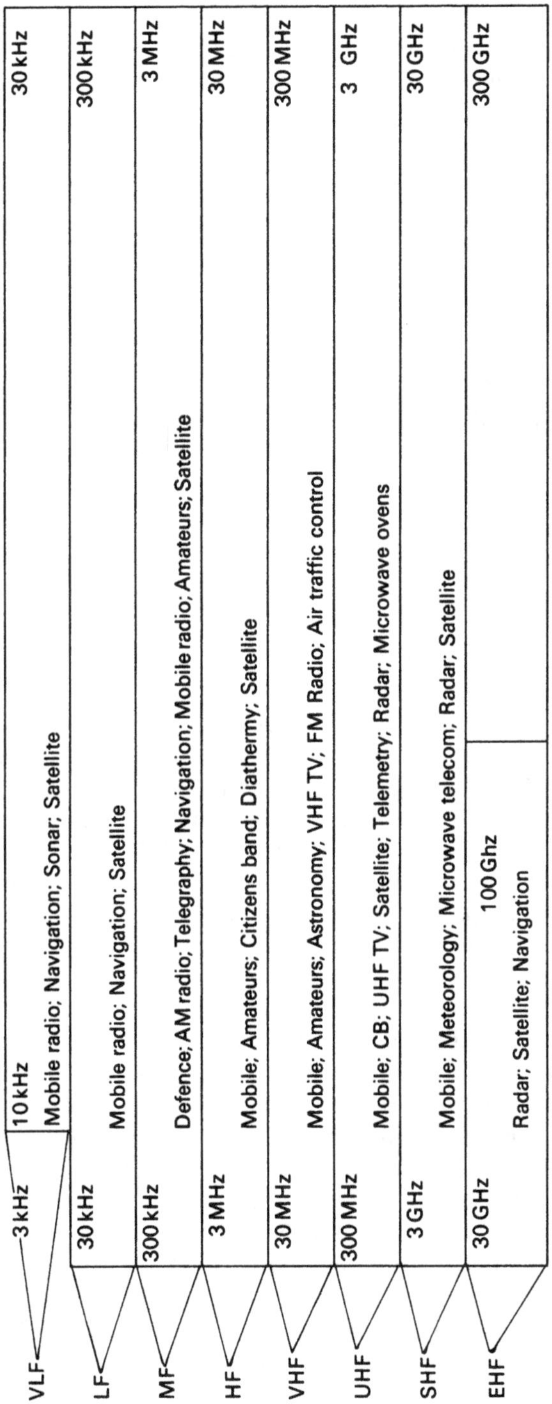

Fig. 16.1 *Radio-frequency spectrum*

long been the useful practice to employ 'quiet' designs of insulator. Elsewhere, with conductor gradients higher by up to 40% than in Great Britain, contributions to the level of interference by the insulators have generally been swamped by conductor corona: improvements in the insulators then become irrelevant[247].

Table 16.1 *Typical requirements for inception, RI and visible corona*

System voltage (kV)	Single unit RI (kV)	Complete string		'Inception' level (dB)*	
		RI	Visual (kV)⊕	Single	String
132	16	110	110	24	12
275	26	240	240	32	18
400	30	320	340	34	22

* Reference level is 1 μV in 37·5 Ω.

⊕ The criterion, in visual tests, is extinction of light from corona, as detected by the dark-adapted eye.

16.2 Effects of capacitance

The distribution of capacitances along an insulator, and their size, govern the electric stresses, which excite generation of interference, and the coupling of the generator to the radiating antenna.

A longrod behaves like a cylinder of dielectric having a relative permittivity about 6. The field intensity falls away rapidly with increasing distance from the live terminal (Fig. 16.2). The generating discharges occur at or near the live terminal, and the capacitance which couples the high-frequency currents into the radiating circuit, i.e. the line and tower, is small. Longrods are thus significantly quieter as interfering sources than strings of discs.

In a string of discs, quite large capacitances – of the order of 30 pF – are connected in cascade through the fittings. The voltage distribution is governed purely by these and by the stray capacitances to line and ground, in dry conditions. In such a voltage-dividing circuit the partition is independent of frequency: identical distributions therefore exist for the power-frequency and for the radio-frequency voltages (Fig. 16.2). The units at the line end are more prone to surface corona than the rest, since their pin-to-cap voltages may be up to three times the average for the string. Insofar as the lower-frequency interference, mainly generated by surface corona, is concerned, these units govern the output of the whole string. (Microspark activity between fittings, on the other hand, may occur randomly and the string may then radiate as its own antenna or as a collection of Hertzian oscillators.)

Because of the high unit capacitances the sources are closely coupled into the line, which presents a load impedance equal to one-half of the line's surge impedance. The circuit constitutes a *CR* network with corresponding frequency attentuation (Fig. 16.3)[248].

In Europe it is common practice to relieve the line end overvoltages by means of stress-grading fittings, which may be corona rings, arcing horns or the bundle of conductors itself. In the USA some lines of 750 kV class cannot employ fittings of this sort because they would impede live working on the line. The designer of the insulators then faces great difficulties in meeting the commonly specified inception voltages and RI limits.

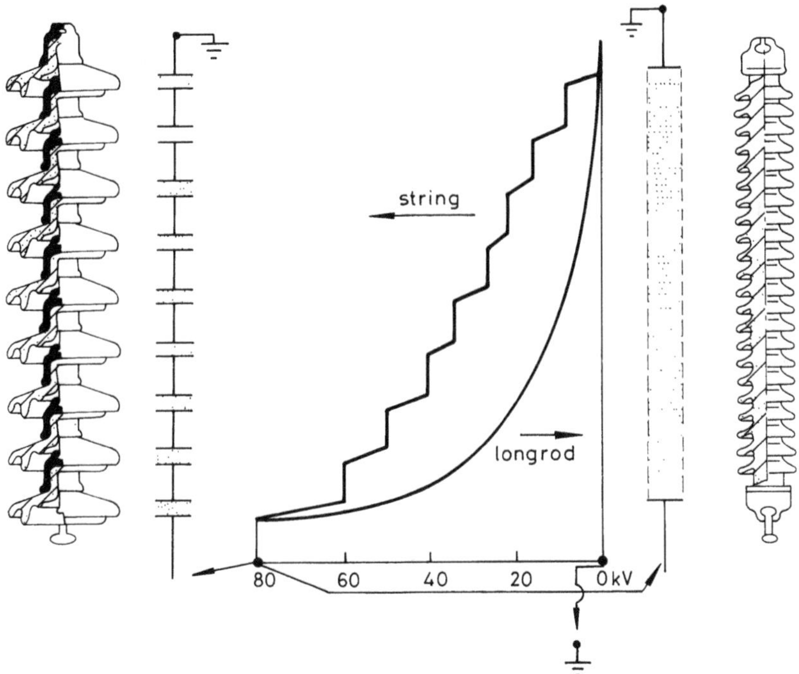

Fig. 16.2 *Voltage distributions along 132 kV string and longrod*
The electrostatic case is shown: clean, dry dielectric: stray capacitances as from normal tower

Some of the devices which are used to minimise surface corona, in cases like these where 'quiet' insulators are essential, are illustrated (Fig. 16.4). Maxstadt[249] showed that gradients between 10 and 14 kV/cm are sufficient to break down air in contact with glass or porcelain over gaps of a few centimetres. The art is therefore to limit the density of displacement current close to the surface of the dielectric by such means as corona-grooves, to pay careful attention to the cement surface and its level, and to interpose subsidiary short skirts, close to the pin cavity, to reduce the stress at the edge of the cement. Unfortunately, some of these devices adversely affect the electromechanical strength of the disc, and all are disadvantageous in some way, not least in increasing costs.

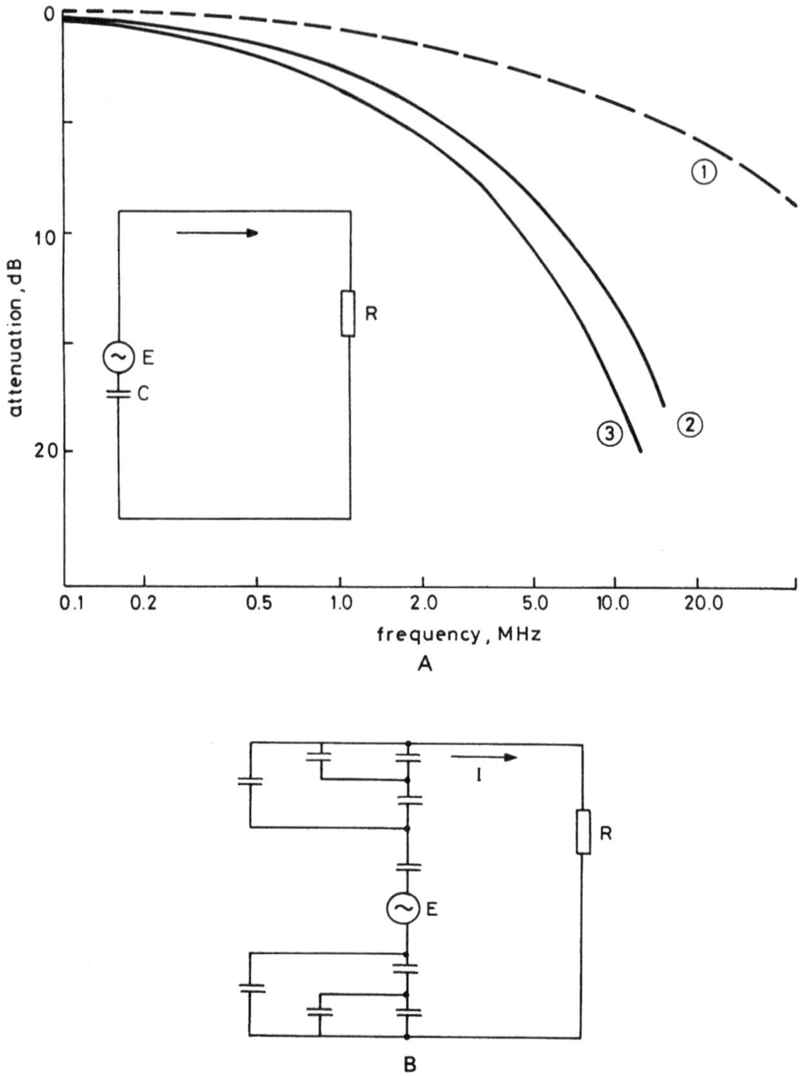

Fig. 16.3 *Interference Generators (Insulators) coupled to line*
A: Single insulator unit
A1: Theoretical curve for $R = 300\,\Omega$, $C = 40\,\text{pF}$
A2: Experimental; medium-inductance test arrangement
A3: Experimental; high-inductance test arrangement
B: String of units, showing stray capacitances to line and ground

16.3 Effects of wetting and pollution

The IEC Standard which governs RI testing of insulators, Publication 437, recommends a clean and dry state for the tested samples. As stated above, these conditions are unlikely to be relevant to real service or to give comprehensive guidance on selection of types. Useful work has been done by Cortina, Rosa and other, mostly Italian, groups[250,251] on likely variations in RI performance, in wetting and pollution, of disc insulators.

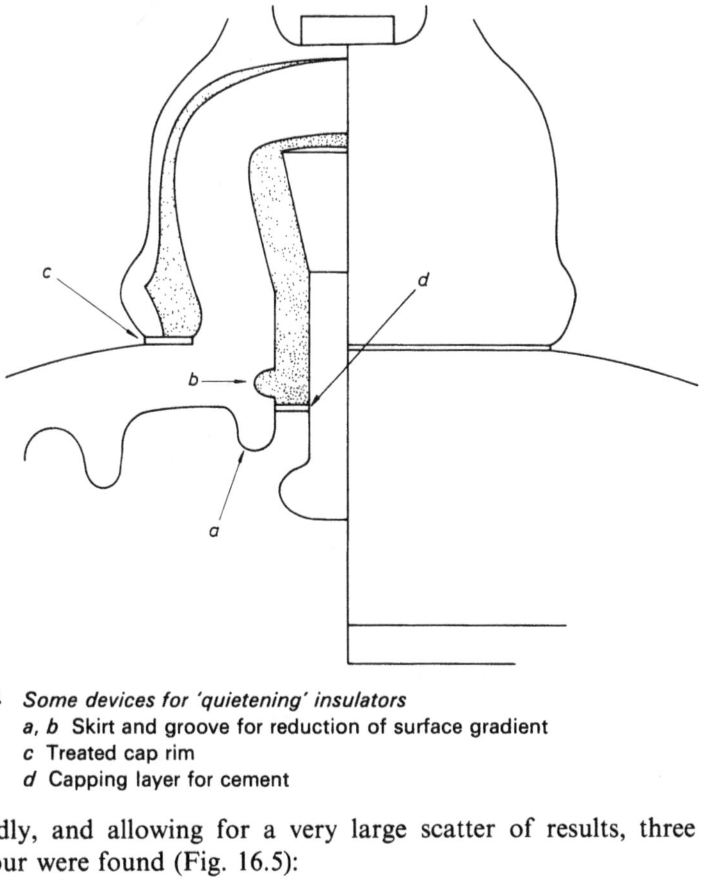

Fig. 16.4 *Some devices for 'quietening' insulators*
a, b Skirt and groove for reduction of surface gradient
c Treated cap rim
d Capping layer for cement

Broadly, and allowing for a very large scatter of results, three types of behaviour were found (Fig. 16.5):

- Cleaned insulators generated less RI as the humidity was increased, the levels tending to constancy above RH 60%.
- Lightly polluted insulators showed the same fall to RH 60%, but this was followed by a rise with RH above some 80%, clearly by onset of a different mechanism.
- Polluted insulators showed no defined trend, with wetting, although other work[252] shows that, as expected, severe pollution causes severe RI.

Cortina's work included a comparison with one type of 'quiet' insulator and led to the conclusion that this particular type, designed for excellent performance in the conditions of IEC 437, was actually noisier than some normal insulators,

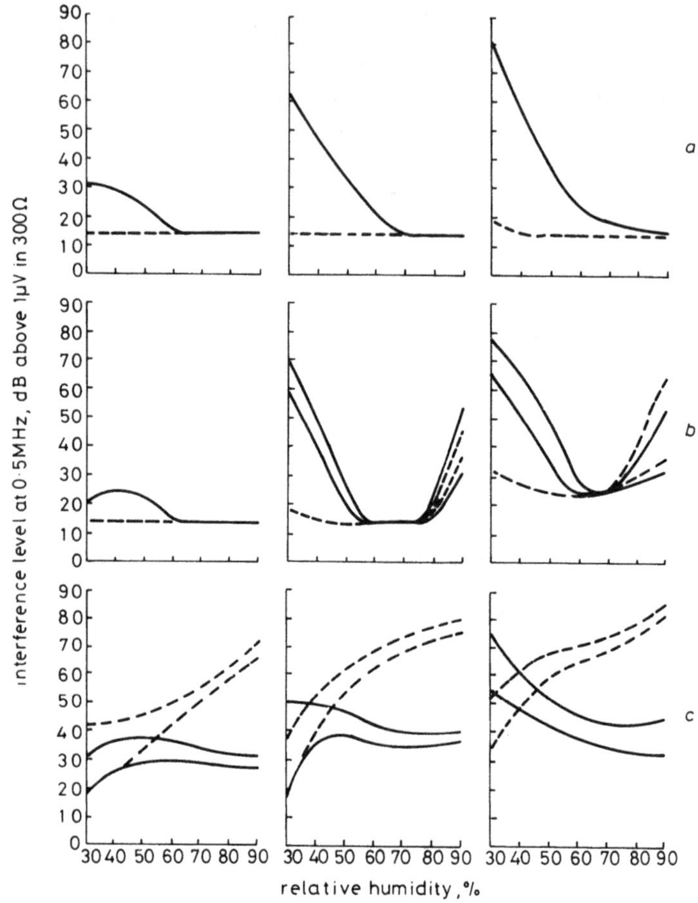

Fig. 16.5 *Variation with humidity of interference level (after References 250 and 251)*
a Carefully cleaned insulators
b Lightly polluted
c Heavily polluted
—— Standard glass cap-and-pin insulator
- - - - Modified 'low-noise' glass insulator

especially in pollution. This, however, is not a general principle: the relative performances will depend on which kind of noise-abating modification is adopted. Resistive glazed insulators, for example, are effectively silent in all atmospheric conditions.

16.4 Acoustic noise from insulators

Electrical discharges on insulators which are wet, polluted or coated with frost cause emission of acoustic noise. This is usually impulsive in nature, a crackling or 'frying' sound modulated at twice the power frequency and sometimes coloured by cavity resonances arising within the insulator's convolutions. Close to the sea, Barber[252] has measured quite high levels of ultrasound in emissions from polluted insulators. Although inaudible to humans, ultrasound has nevertheless caused nuisance by provoking persistent barking in kennels located close to lines (lightheartedly called the 'dog amplifier' effect!).

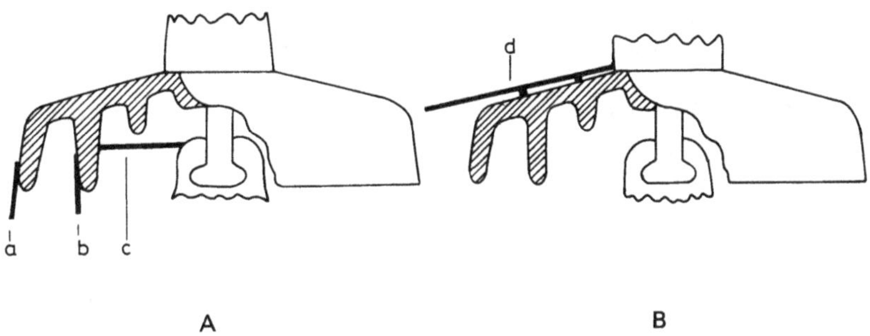

Fig. 16.6 *Suppressors for aerodynamic howling*
A. Either (*a*) or (*b*), but not both together, act to spoil the process of excitation
The resonant cavity is detuned and damped by (*c*)
B. Booster shed acts both to spoil and detune (*d*)
About one in four units needs treatment

A rare but objectionable form of AN from insulators is howling, induced aerodynamically. This arises only with certain profiles which include cavities between long skirts terminating in a plane (Fig. 16.6) and in a range of wind speeds of about $18 - 22\,\mathrm{m\,s^{-1}}$. The mechanism is a cooperative one, the level of oscillation rising with the number of discs in the string, and the emitted tone is pure, having a bandwidth of only a few Hz, and lying in a part of the audible spectrum, a few hundred Hz, to which the human ear is very sensitive (Fig. 16.7).

It is known that the substitution of a different shape of insulator for every third or fouth unit in a noisy string will suppress the howl[253,254], but mixed strings of this type are electrically and economically undesirable. Tunstall and the author have shown, however[255], that suppression may also be effected by fitting polymeric parts to modify the air flow over some one in four units, and that Booster sheds, which are electrically beneficial, act well as silencers (Fig. 16.6).

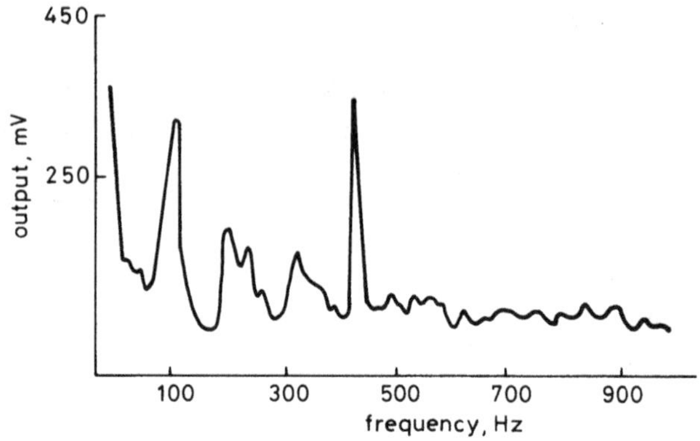

Fig. 16.7 *Insulator howl*
Acoustic spectrum from six equal-skirted units: peak is at 464 Hz. Low-frequency peaks are wind-tunnel noise. Measured bandwidth is below 1 Hz

These are easily fitted in minimum outage time, or even live if necessary, and their use can be confined to inhabited localities from which complaints are likely.

Chapter 17

Insulator of the future

17.1 Indicators from known facts

The strong similarity between most of the insulators which are being installed today and those which were current in 1912 shows not so much a lack of innovative skills among electrical engineers as the remarkable intelligence and percipience of the early pioneers. A correct first guess leaves little scope for subsequent improvement.

It is, however, useful to consider some of the conceptions, over the past nine decades in the world of insulators, which have appeared promising. Many have either aborted or withered in infancy: these provide a salutary background for some of today's optimistic offerings of insulators having exceptional performances.

Since it is the vulnerability of surfaces to water and pollution which mainly limits the performance of insulators, the perfect surface, having, in Gavey's words as long ago as 1878, 'if possible infinite resistance' has long been the Holy Grail of the researcher. Gavey recognised that the oil bath would provide such a surface; mobile pastes and greases have approached it; solid hydrophobes have fallen short of it. The oil bath has been defeated by the practical matter of spillage in wind ; greases have become choked or have deteriorated; solids have lost their non-adhesive and non-wetting virtues, under the attacks of weather, blanketing and discharges.

The second generation of surface improvers has been based on the assumption that some degree of contamination is inevitable, and has aimed to suppress the consequent discharges or limit their damaging effects. Here are comprised the active fillers in polymeric housings, which have extinguished discharges by emitting vapour, and have catalytically purged carbon arising from destruction of the polymer by heat from the discharge.

The greatest benefactor of all surface treatments has, however, been the resistive glazes. These, the antithesis of 'infinite resistance' dielectrics, have conferred large improvments in all the relevant properties, flashover voltage, absence of electromagnetic interference, freedom from acoustic noise, at a modest cost in lost energy.

On 'miraculous shapes' for insulators Gavey must again be admired for his choice. Of the 29 shapes shown in his paper he favoured the 'double cup' form, not much different from today's anti-fog profile. Since his day, would-be improvers have first sought to keep parts of their insulators dry by shields or umbrellas, wrongly assuming that rainwater was an enemy instead of a friendly remover of dirt. Later they have chosen the opposite tack and sought full exposure with minimum aerodynamic drag: here may be mentioned the double-convex aerofoil, invented by the author and a resounding failure under test, and some 'aeroform' shapes of Motor and longrod insulator, similarly disappointing under test or in service.

The helicoidal shedded post was another concept which seemed a good contender on the drawing board but turned out a failure in the field. The mental picture of rainwater flowing helter skelter round the helix, to leave a clean and insulating barrier, was a gift to the salesman; but the product was a disaster. One design for 110 kV gave good performance in the author's Brighton Testing Station – as a support for samples to run no higher than 11 kV! (Fig. 13.6 *i* and *j*).

Sloping planar sheds provided another interesting choice of shape, first, some 50 years ago, on discs and more recently on the TDL polymeric longrod, again one of the author's involvements. The good draining and aerodynamic properties have proved beneficial, with some difficulties in torrential wetting. This design has fallen from favour because of materials' shortcomings, not because of its shape (Fig. 17.1).

Radical changes in shapes, as well as in almost all other properties including strength for given weight and freedom from impact damage, arose from the introduction of the fibrous-cored polymeric insulator and its cousin (Fig. 17.2), the fibrous-cored ceramic (also known confusingly as the composite or unitary) longrod. From the earliest times of Ash and Dey's innovation in 1948, the fibrous-cored insulator has been 'on the verge' of driving the classical insulator into obsolescence. Minor problems have turned into major problems, once solved then followed by others – as one embittered researcher remarked, like battling to the peak of a mountain range to find not the open sea but another range of mountains to be scaled – until today when, it seems, the reliable fibrous core may be as practically attainable as the frictionless bearing.

The notion of the concrete insulator, a structure which ideally could be cast on site and would safely withstand both large mechanical loads and high voltages, has long had a strong appeal. The Polysil, and recently the NIM materials, have been steps in this direction, while a porous insulator, of cement designed to sweat oil continuously, was tried unsuccessfully by the author's group in the 1960s. A related product, the strippable compounds developed by O. Pordes about the same time and guided by the release agents used in concrete

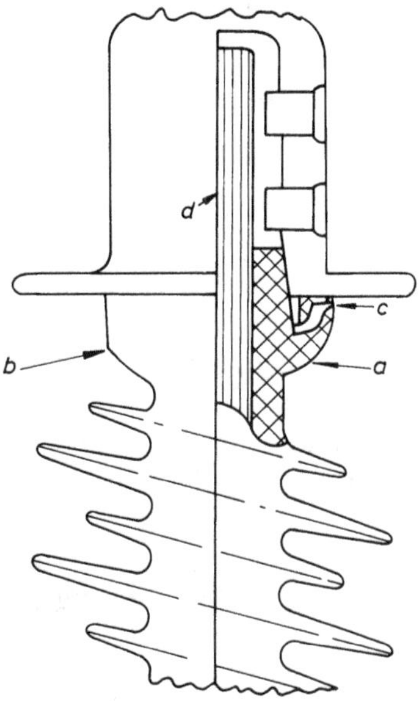

Fig. 17.1 *Sloping planar shedded epoxy longrod*
a One-piece epoxy casting
b Enlarged terminal section of epoxy
c Silicone elastomer seal
d Strong laminated core

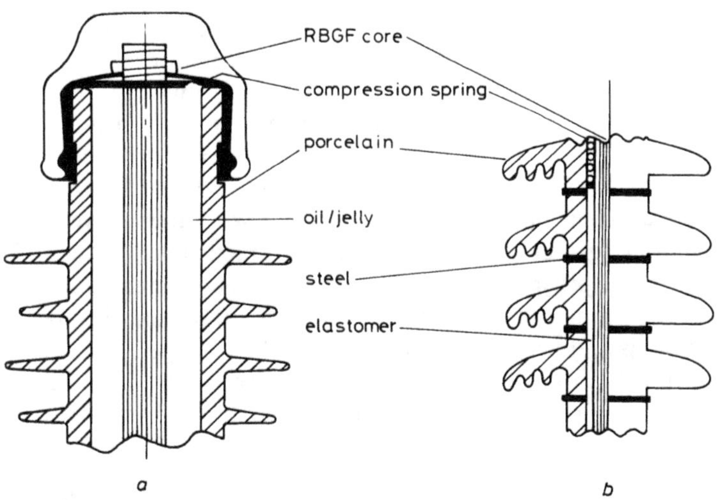

Fig. 17.2 *Porcelain-housed RBGF-cored composite insulators*
a Dimex principle; impregnated wooden core, oil or jelly filling
b Joslyn principle: steel-spaced porcelain elements, greased core, compressed elastomer packing

casting, was an outstanding surface treatment, condemned for shortcomings which were probably non-essential.

Separation of the mechanical from the electrical functions of an insulator, first employed in the fibrous cored polymeric types, has since been made the basis of the Hybrid insulator, in which porcelain carries the loads and polymer controls the leakage current. The other functions of the profile, such as regulating the drainage rate under wetting, blocking flashover caused by droplet motion or drips, and determining the aerodynamic flow patterns which are responsible for the density and locations of surface deposits, are now being handled separately by various kinds of addition. These include Booster sheds, creepage extenders and edge protectors, all made of polymers generally having very different properties from those of the substrate, and consequently liable to introduce changes in behaviour additional to the ones intended and not always beneficial.

Finally may be mentioned the surprising results which have followed reduction of the insulator diameter to the millimetre range, as in the monofilaments, and of the forced concentration of leakage current at a multiplicity of contact points, as in the chains. Remarkably good electrical potentialities are, at present, limited by mechanical weakness.

17.2 Extrapolations from current practices

Where high costs arise from interruption of supply following flashover of insulators, or where there are incidental disadvantages – for example, busbar flashover at a coastal nuclear station may lose high-merit generation and also cause reactor poisoning – the ideal insulator is no insulator. Enclosed, metalclad substations are there sometimes used in which the only exposed surfaces are those of the cable sealing ends or bushings. These can be protected by frequent live washing (Fig. 1.2). At the time of writing, however, metalclad apparatus itself is suffering from flashover of a different character across internal insulation, and the designer of the plant is thus faced with alternative risks.

Although underground transmission is now practicable for large-load transfer capabilities, the prospect of its ousting overhead lines seems still to be remote for both economic and environmental reasons, other than in areas of dense population or special need. The million-volt, UHV, range of insulator and HV DC types, already in use on a small scale, are nearly certain to be widespread at the start of the next century. Acceptable flashover rates with such high-powered lines will be small: all the design parameters, material, shape and mechanical system will have to be optimised, and costs are likely to rise.

A reasonable guess, on the subject of future insulators for UHV, is that the surface will be polymeric. The possible permutations of organic material and filler are so numerous, and the scope for designing-in specific properties is so wide, that ceramics and glass, whose sole virtue is longevity, are likely to be superseded. This is not to say, however, that their mechanical role will also disappear; on the contrary, the prospects for special ceramics, using the new toughening ingredients, for glasses which are surface treated, devitrified to yield strong glass-ceramics, or both, and even for the cement-based mineral bodies are seen as highly promising[256,259]. Insofar as HV DC presents special problems, both of contamination processes and creepage requirement, insulators for overhead lines will probably be about the same as for UHV AC., but substation insulators will almost certainly differ. Very long profiles, with added barriers against water both as cascades and ejected droplets, seem a likely solution to an intrinsically difficult problem.

The mechanical demands on overhead lines will include loads of 100 tonne order, for which disc insulators are a clumsy answer. It was with this class of load in mind that so much work was put into the use of laminated fibrous cores, as against pultruded ones[49]. At present it seems likely, in view of the excellent mechanical record of laminated materials and their apparent immunity to the troubles which dog pultrusions in conditions of combined chemical and other stresses, that laminated cores will come back into favour. An added virtue is that such materials can be tailored, with different ratios of cross and axial filaments at the ends from those in the main body, to accept mechanical couplings which give high reliability[49].

For high voltages and long insulators the common practice of hanging lines below crossarms is illogical, since the length of the string must add to the height of the tower: costs of towers and foundations rise very rapidly with increasing height, as do the numbers of complaints from the environmental lobby. Insulating crossarms or other mechanical designs, in which support comes from beneath or from the side, are therefore always attractive. In the early days of polymer–concrete research it was envisaged that structures might be developed in which both electrical and mechanical strengths would be high enough to permit direct attachment (Fig. 9.10). This possibility may well materialise, although the cognate dream of a system which could be poured and cured in the field, like a conventional cast-concrete structure, seems doomed by the processing requirements of the most promising new concretes, which currently comprise extrusion and curing in an autoclave.

Composite materials using strong polyamide fibres instead of glass have already been successfully used in cross-braced structures which are electrically insulating. These are intrinsically less vulnerable than RBGF to internal propagation of water, especially when the fibres can be laid non-parallel to the electric field. The ingenious Metapol insulator (Fig. 17.3) eliminates the fibre core; however, other materials' difficulties then arise. Helical fibrous cores can confer both long creepage path and controlled elasticity (Fig. 17.3).

Moving into the world of fantasy, a spider's web assembly could utilise the remarkable properties of monofilaments in resisting pollution flashover. Attempts have already been made to copy the human eyelid's function of wiping clean the eyeball, by adding a moving wiper to the top surface of an insulator.

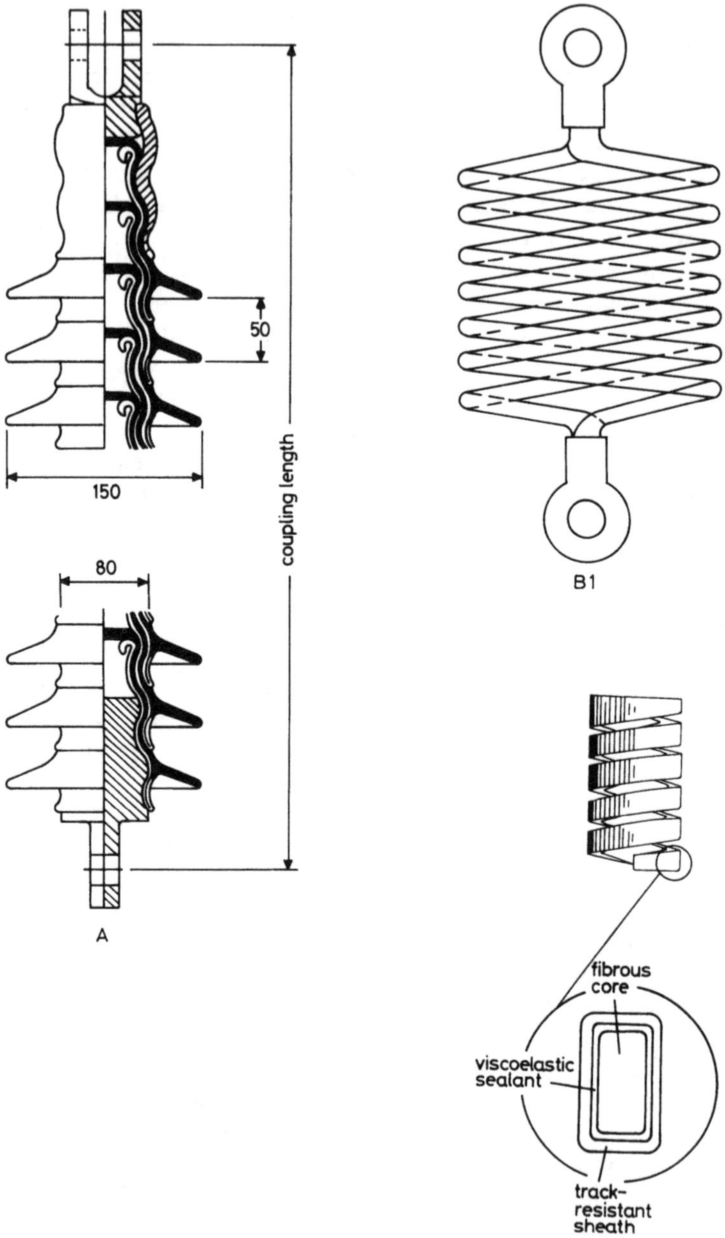

Fig. 17.3 *Glasson's Metapol insulator: Helical insulators*
Dimensions in mm
A. Metapol design: swaged metal and polyolefine with no fibrous core, (Dulmison Company).
B. Helical 'spring' insulators: (1) USSR design: (2) Raychem Co.

There seems more promise, however, in the use of 'eyelashes', to catch pollution before it impacts upon the insulator: arrays of parallel filaments are effective particle traps (Fig. 17.4).

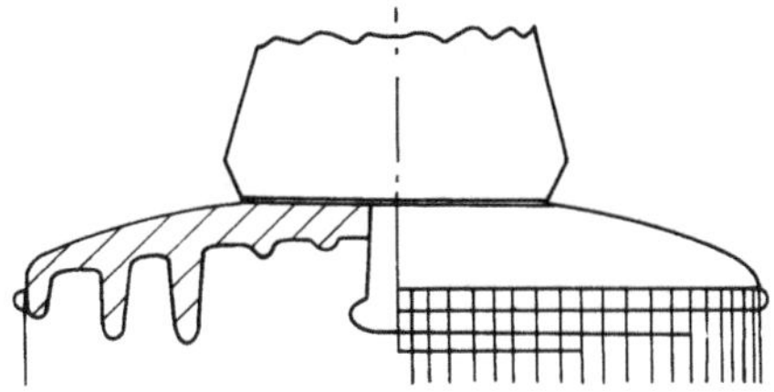

Fig. 17.4 *'Eyelash' insulator*
The array of insulating monofilaments protects the creepage from contamination

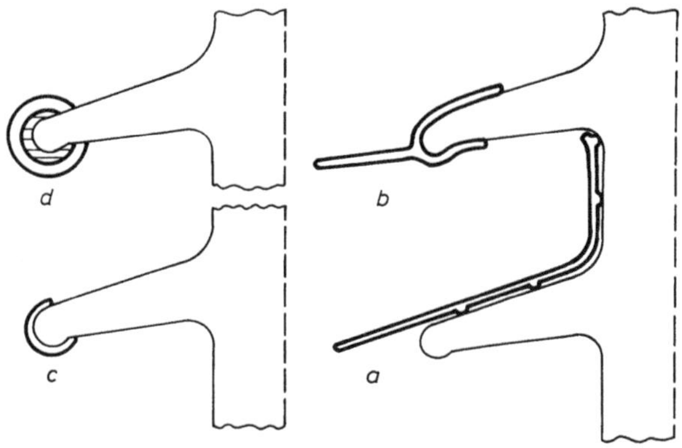

Fig. 17.5 *Current types of polymeric addition*
a Booster shed
b Creepage extender
c Buffer
d Stress reducer, or dry collar
The Salmon vandal shield is similar to *a*, but uses ribs instead of nibs, or is unspaced

Insofar as open-type substations are concerned there is no doubt that scope exists for radically better insulating systems than are currently available, and that the need will increase with rising voltage and adoption of DC. Although very large bushing shells are already made, using the technique of glaze jointing, and multi-section hollow shapes can be joined successfully by means of organic adhesives, the electrical performances are poor because of the large diameters, costs are high, and vulnerability to mechanical shocks, especially earthquakes[260], is a cause for concern.

There seems no difficulty, in principle, in using braced structures based on fibrous composite or other ultra-strong cores, in both cases with polymeric sheaths, for substation duties like busbar support. Rather it is the housings for circuit breakers, current and voltage transformers, as well as the bushing shells, which are intrinsically difficult. Some functions are likely to disappear, for example where optically linked transducers supplant conventional information conduits or when air-blast switchgear finally dies out; but large cylinders of one sort or another will inevitably be with us as long as high-voltage lines terminate somewhere.

Polymeric additions to deal with water and pollution seem highly likely to proliferate (Fig. 17.5), but the opportunities for some sort of *cordon sanitaire* of precipitators to remove contaminant dusts and droplets, or of air-blasts to modify the natural aerodynamics, will evidently increase as the costs of flash-overs move towards the unacceptable.

References

1 GAVEY, J.: 'Insulators for aerial telegraph lines', *J. Soc. Tel. Engrs.*, 1878, **7**, (22) & (23)

2 MITTLER, O.: '75 years Brown Boveri', 1966, p. 15

3 LUNDQUIST, E. E.: 'Transmission line construction' (McGraw-Hill, 1912)

4 BRENT MILLS.: 'Porcelain insulators and how they grew' (Brent Mills, 1970)

5 CHERNEY, E. A.: 'Cement growth failure of porcelain suspension insulators'. AMEU, Ottawa, 1982

6 WEICKER, W.: 'Einzeldarstellungen aus der elektrotechnik' (Julius Springer, Berlin, 1932)

7 ELY, C. H. A., and LAMBETH, P. J.: 'Artificial pollution test for high voltage outdoor insulators'. *Proc. IEE*, 1964, **111**, pp. 991–998

8 FUJIMURA, T., NAITO, K., and IRIE, T.: 'Performance of semiconducting glaze insulators under adverse conditions'. *IEEE Trans.*, 1978, **PAS-97**, pp. 763–771

9 NIGOL, O., REICHMAN, J., and ROSENBLATT, G.: 'Development of new semiconductive glaze insulator', *IEEE Trans.*, 1974, **PAS-93**, pp. 614–621

10 SMITH, E. J. D.: 'The corrosion of semiconducting glazes', *Trans. Brit. Ceram. Soc.*, 1959, **58**, pp. 277–300

11 HOGG, P. M.: 'The development of a pre-stressed (toughened) glass insulator', *J. IEE*, **87**, 1940 p. 615

12 JOHNSTON, E. F.: 'Toughened glass insulators', *Elect. Rev.*, 19 Feb. 1960

13 CLABBURN, R. J. T., and LOOMS, J. S. T.: 'A hybrid system for outdoor insulation', IEE DMMA Conf., 1984

14 PEIXOTO, C. A. O., PARGAMIN, L., MARRONE, G., and CARRARA, G.: 'Failure of transmission line cap and pin ceramic insulators under dc stresses'. CIGRE Study Cttee. No. 36-03, 1983

15 LECOMTE, D., and MEYERE, P.; 'Evolution of design for 753 kV lines'. CIGRE 22-08, 1980

16 KARADY, G., and LIMMER, H.D.: 'Design and application of high-voltage synthetic insulators'. American Power Conference, 1979, 19**B**

17 LIFSHITZ, J. M.: 'Time-dependent fracture of fibrous composites', *Composite Materials*, 1974, **5**; and 'Fracture and fatigue' (Academic Press, 1974)

18 WATTS, A. S.: 'American and European insulators'. *Elect. World*, 1910, **55**, p. 1294

19 WEIBULL, W.: 'Statistics of strength', *Ing. Vetensk. Akad. Proc.*, 1939, pp. 151, 153; and *J. Appl. Mech.* 1951, **18**, p. 293

20 KINGERY, W. D., BOWEN, H. K., and UHLMANN, D. R.: 'Introduction to ceramics' (J. Wiley, 1976) 2nd Edn. pp. 786–789

21 IEC Publication 233: 1974 'Tests on large hollow porcelains for use in electrical installations'

22 STANDRING, W. G.: 'The electric strength of solid dielectrics under impulse voltages', *J. IEE*, 1941, **88**, Pt. II, pp. 360–365

23 GIBSON, H., and BALLARD, J. G.: 'Some effects of long term DC stresses on porcelain operated at high temperature', *J. Mat. Sci.*, 1972, 7, pp. 303–315

24 SKOIBEDO, A. I., TIKHODEEV, N. N., ZHUKOV, V. V. and NIKOLAEV, N. A.: 'On the use of tempered glass insulators on DC overhead transmission lines', *Soviet Power Enging.*, 1972, pp. 310–313

25 FIERO, D. C., DEWEY, B. F., and CROCKER, J. E.: 'Flaw and fracture phenomena in electrical porcelain', *IEEE Trans.*, 1970, **PAS-89**, pp. 681–683

26 VALEEV, Kh. S., *et al.*: 'Aging of suspension insulators under high voltage', *Elekt. Stanstii*, 1966, **4**

27 BLOOR, E. C.: 'The composition and properties of electrical ceramics' in 'Ceramics – A symposium' (British Ceramic Society, 1953) p. 250

28 PAUL, A.: 'Chemistry of glasses' (Chapman & Hall, 1982) p. 1

29 ZACHARIASEN, W. H.: 'The atomic arrangement in glass', *J. Am. Chem. Soc.*, 1932, **54**, p. 3841

30 ANGUS, H. T.: 'Physical and engineering properties of cast iron' (British Cast Iron Research Assoc., 1960)

31 MARSH, J. D., and SWIFT, D. A.: CERL Note, 1977 (see Reference 48)

32 MOREY, G. W.: 'The properties of glass' (Reinhold Publishing Corp. NY, 1954) p. 113

33 BRADWELL, A., and WHEELER, J. C. G.: 'Tensile failure of glass fibre'. Reinforced Plastics Congress, Brighton, 1982, p. 121

34 OTTO, W. H.: 'E-glass fibers'. Proc. 6th Sagamore Ord. Mat. Res. Conf., 1959, p. 277

35 COJAN, M., PERRET, R., MALAGUTI, C., NICOLINI, P., LOOMS, J. S. T., and STANNETT, A. W.: 'Polymeric transmission insulators'. CIGRE 22-10, 1980

36 CHANDLER, H. D., JONES, R. L., and REYNDERS, J. P.: 'Stress corrosion failure of composite longrod insulator'. 4th ISH, Athens, 1983, 23.09

37 OWEN, M. J., HARRIS, S. J., and NOBLE, B.: 'Failure of electrical insulators with pultruded cores', *Composites*, 1986, **17**, pp. 217–226

38 HOEVEL, A.: 'Improved reliability of fibre optic overhead cables'. Proc. Internat. Wire & Cable Symposium, 1986

39 FINK, M. H.: 'Electrical insulating rod'. US Pat. 2 997 529, 1961

40 LIFSHITZ, J. M., and ROTEM, A.: 'Time dependent failure of fibrous composites', *Fibre Sci. Technol.*, 1970, **3**, p. 1

41 ISHAI, O., and MAZOR, A.: 'The effect of environmental loading history of longitudinal strength of glass-fiber reinforced plastics', *Rheol. Acta.* 1974, **13**, pp. 381–394

42 ROMANENKOV, I. G.: *Soviet Plastics*, 1967, **4**, pp. 43–45

43 CURTIS, H. L.: 'Surface and volume resistivity', *Bull. Bur. Standards*, 1914, **11**, p. 359

44 VIBRANS, G. E.: 'Vacuum voltage breakdown'. MIT Lincoln Lab. Tech. Rep. 353., 1964

45 RIDOUT, K., and LOOMS, J. S. T.: 'Artificial faults'. Reported in Reference 67

46 PENNECK, R. J.: 'Track resistant polymer'. Raychem internal report, 1970. See MORTON-JONES, D. H., and ELLIS, J.W.: 'Polymer products': (Chapman & Hall, 1986) Fig. 26.7

47 SCARISBRICK, R. M.: 'Electrochemical erosion of epoxy-resin outdoor insulation', *Proc. IEE*, 1974, **121**, pp. 779–783

48 ELY, C. H. A., LAMBETH, P. J., LOOMS, J. S. T., and SWIFT, D. A.: 'Discharges over wet, polluted polymers'. CIGRE. 15-02, 1978, pp. 4–5

49 LOOMS, J. S. T., and PROCTOR, F. H.: 'The development of an epoxy-based insulator for UHV', IEEE A 76 342–6, 1976

50 CLABBURN, R. J. T., PENNECK, R. J., and SWINMURN, C. J.: 'Some parameters affecting the outdoor performance of plastic materials'. IEEE C 72 501–5, 1972

51 PARR, D. J., and SCARISBRICK, R. M.: 'Performance of synthetic insulating materials under polluted conditions', *Proc. IEE*, 1965, **112**, pp. 1625–1632

52 FREY, A. M., and PERRY, E. R.: 'Epoxy post type voltage graded insulator', IEEE 31 CP 67–454, 1967

53 RYSHKEWITCH, E.: 'Ceramic tools', *J. Amer. Ceram Soc.*, 1953, **36**, p. 65

54 SINGER, F., and SINGER, S. S.: 'Industrial ceramics' (Chapman and Hall, 1963) pp. 741–746

55 PARMELEE, C. W.: 'Ceramic glazes' (CBI Publishing Co., Boston, Mass, 1973) 3rd. Edn., pp. 242–243
56 HYSLOP, J. F.: 'The action of heat on clays' *in* 'Ceramics – a symposium' (British Ceramic Soc., 1953) pp. 186–200
57 BRINDLEY, G. W., and MacEWAN, D. M. C.: 'Structural aspects of the mineralogy of clays and related structures', *ibid*; Ref. 3.4, pp. 15–55
58 DUMORA, D., PARGAMIN, L., and PARRAUD, R.: 'High voltage suspension ceramic insulators' (CERAVER, St. Yorre, France, 1984)
59 PENTELOW, A., and PROCTOR, F. H.: 'Anvil tests on RBGF rods subjected to acid attack'. TDL/Permali Tech. Rep. (TDL/Permali, Bristol Rd., Gloucester, 1978)
60 CIGRE: 'Composite Insulators'. Study Cttee. 22, WG 22-10, 1980
61 CHANDLER, H. D., and REYNDERS, J. P.: 'Electro-chemical damage to composite insulators': CIGRE, 22-10, 1984
62 AKHTAR, A., and de TOURREIL, C.: 'Mechanical behavior of composite insulators'. Canadian Elect. Assoc. Symposium, Montreal, 1985
63 BRADWELL, A., and WHEELER, J. C. G.: 'Evaluation of plastics insulators for use on BR 25 kV overhead line electrification'. *Proc. IEE*, 1982, **129B**, pp. 101–110
64 RAYCHEM Ltd.: 'Polymeric and hybrid insulators'. Catalogue, 1986
65 FURSE, W. J. & Co.: 'The Hetherlite insulator'. Catalogue, 1984
66 ARMSTRONG, J. A.: 'Service experience on Ohio Brass Hi-Lite insulators'. Canadian Elec. Assoc., Vancouver, 1983
67 PROCTOR, F. H.: 'Polymeric insulators', *Middle East Electricity*, March 1984, pp. 37–39
68 CIGRE: Paper 3 to Study Cttee. SC 22-03, 1978
69 CIGRE: Paper 23, IWD 10, to Study Cttee. SC 22-10, 1980
70 PARGAMIN, L.: 'A polymeric insulator for 765 kV class' (CERAVER, St. Yorre, France, 1976)
71 LOOMS, J. S. T., *et al.*: 'Cast epoxy systems assessed for HV. systems', *Elect. Rev.*, 12 May 1976, pp. 663–664
72 LOOMS, J. S. T.: 'Insulator shapes for the 1980s', *Elect. Times*, 14 March 1975
73 LUCAS, D. H.: 'New approach to outdoor insulator design', Central Electricity Res. Labs., RD/L/R 1391, 1966
74 LUCAS, D. H.: 'Design of an everclean window'. Brit. Pat 5654/59, 1959 (with P. A. CROSS and W. A. SNOWSILL)
75 ELY, C. H. A., KINGSTON, R. G., and LAMBETH, P. J.: 'Artificial and natural pollution tests on outdoor 400 kV substation insulators', *Proc. IEE*, 1971, **118**, pp. 99–109
76 CLABBURN, R. J. T., and LOOMS, J. S. T.: 'Developments in hybrid outdoor insulators'. BEAMA Conf., Brighton, 1986
77 BAUMANN, D., MUNK, K., and REMBOLD, H.: 'The erosion resistance of epoxy casting systems'. 3rd ISH, Milan, 1979, 23.1
78 LOOMS, J. S. T., and ECCLESTONE, B. G.: 'Discussion on Ref. 63'. *IEE Proc.*, 1983, **130B**, pp. 127–133
79 LAMBETH, P. J., LOOMS, J. S. T., ROBERTS, W. J., and DRINKWATER, B. J.: 'Natural pollution testing of insulators for UHV transmission systems'. CIGRE, 33-12, 1974, pp. 4–6
80 BRADWELL, A., and WHEELER, J. C. G.: 'The use of plastic insulators on British railways'. DMMA, Brighton, 1979, pp. 330–332
81 BRADWELL, A., and WHEELER, J. C. G.: 'Evaluation of plastic insulators for BR 25 kV electrification'. *IEE Proc.*, 1982, **129B**, pp. 101–110
82 BAUER, E., KAERNER, H., MUELLER, K. H., and VERMA, M.P.: 'Service experience with the German composite longrod insulator'. CIGRE, 22-11, 1980, pp. 13–15
83 ELY, C. H. A., and LOOMS, J. S. T.: 'Shrunk on sheds for live line tools', *Elect. Times*, 1968, **154**, p. 245
84 COPPACK, J. L., and GIBBON, W.: 'Method of attaching fittings to rods or tubes of RBGF'. British Patent to BICC Ltd., 816 926, 1957; and BANNERMAN, N.G.: 'Brit. Pat. 902 197, 1959

85 KAMINKSI, J.: 'Long term mechanical and electrical strength of suspension insulators', IEEE Trans., 1963, pp. 446–451

86 OHIO BRASS CO.: 'Suspension insulator mechanics', *Hi-Tension News*, 1963, **32**, p. 11

87 LAPP INSULATOR Co.: 'Suspension insulators'. Catalog 9-B, 1970, pp. 2–3

88 NOZAKI, H., and INOUE, H.: 'High tensile strength suspension insulators'. US Patent to NGK Ltd., 4 396 798, 1983

89 TAYLOR, D.: 'Disc insulators: a technical review and a design philosophy' (Doulton Insulators Ltd., Tamworth, UK, 1981)

90 MIER-MAZA, R., LANTEIGNE, J., and de TOURREIL, C.: 'Failure analysis of synthetic insulators'. IEEE 83 WM 133–6, 1983

91 PENTELOW, A.: 'Comparative strengths of cylindrical, waisted and fluted RBGF terminations'. TDL/Permali, Gloucester, England, 1980 (unpublished)

92 SINGER, F., and SINGER, S. S.; 'Industrial ceramics' (Chapman & Hall, 1963) pp. 349–350

93 American National Standard C.29-2. 1977

94 SALMON, R. G.: 'Vandal shields for insulators', *Distribution Developments*, June 1977, pp. 19–25, and March 1979

95 CHERNEY, E. A.: 'Cement growth failure of porcelain suspension insulators'. AMEU Meeting, Ottawa, 1982

96 CEGB South Western Region: 1981. In course of publication

97 PROCTOR, F. H.: 'Insulators and sand storms', *Middle East Elect.*, Feb. 1977

98 FGH, MANNHEIM: 'Arc tests on Raychem insulator', (Raychem Ltd., Swindon, UK, 1975)

99 COJAN, M., PUCK, A., and RIVIERE, D.: 'Reduction of the dimensions of towers by means of insulating elements of synthetic material'. CIGRE, 22-10, 1974

100 LAPP INSULATOR CO.: 'Line post insulators'. Catalog 10-A, 1977, p. 32 (hinge), p. 38 (load limiter)

101 BRADWELL, A.: 'Importance of preventing moisture ingress to polymeric insulators', *IEE Proc.*, **131B**, 1984, pp. 245–251

102 DEWEY, B. F., and FALTER, S. L.: 'Tests on insulators reveal danger signs for silicone', *Elect. World*, 1969, p. 25

103 GERTSIK, A. K., KORSUNTSER, A. V., and NIKOLSKII, N. K.: 'The effect of fouling on insulators for h.v.d.c. overhead lines', *Direct Current*, 1957, **3**, pp. 219–226

104 WALSHE, D. E. J.: 'Wind tunnel experiments on the pollution of suspended high voltage insulators'. National Physical Laboratory, Teddington, England, Aero Rep. 1083, 1963

105 LAPP INSULATOR CO.: 'Desert design insulators'. Catalogue, 1980

106 LAST, F. H., PEGG, T. H., SELLERS, N., STALEWSKI, A. C., and WHITTAKER, E. B.: 'Live washing of h.v. insulators', *Proc. IEE*, 1966, *113*, pp. 847–860

107 YASUDA, M., and FUJIMURA, T.: 'High water pressure hot line insulator washing equipment for 500 kV substations', *IEEE Trans.*, 1976, **PAS-95**, pp. 1919–1928

108 PENNECK, R. J.: Private communication. Raychem Ltd., Swindon, UK

109 EL-ARABATY, A., NASSER, E., EL-SARKY, A., *et al.*: 'Selection of insulators for . . . -desert conditions'. CIGRE, 33-11, 1980

110 HALL, J., and ORBECK, T.: 'RTV protective coating for porcelain insulators'. IEEE. 82 SM 363-0, 1982

111 A. B. CHANCE DIVISION, EMERSON ELECTRIC.: 'Remedies for insulator pollution'. Catalogue, Portishead, UK, 1986

112 ATKINS, A. D., and CLABBURN, R. J. T.: 'Insulators for railways in the Indian Subcontinent'. Indian Railways Conf., 1985

113 NGK INSULATORS LTD.: 'Study of pollution problem'. Technical Note 68069, Mizuho, Nagoya, Japan, 1968

114 LAMBETH, P. J., LOOMS, J. S. T., SFORZINI, M., CORTINA, R., PORCHERON, Y., and CLAVERIE, P.: 'The salt fog test and its use in insulator selection', *IEEE Trans.*, 1973, **PAS-92**, p. 1884

115 LAMBETH, P. J., AUXEL, H., and VERMA, M. P.: 'Methods of measuring the severity of natural pollution as it affects HV insulator performance', *Electra*, 1972, p. 20

116 CERL.: 1982. Internal report

117 PETRUSCH, W., LANGE, G., SCHMITT, W., KLUGE, W., and SCHUMANN, P.: 'Experiences on the continuous monitoring of the leakage current on polluted insulators'. CIGRE, Stockholm Symposium 22-81, 122-03, 1981

118 VERMA, M. P.: 'Hochster ableit stromimpuls als kenngroesse fuer das isoliervernalten verschmutzter isolatoren', *ETZ-A*, 1973, *94*, pp. 302–303

119 STALEWSKI, A. C.: 'Automatic pollution detector and washing system for insulators', *CEGB Disclosure Bull.*, 1969, No. 122

120 SFORZINI, M., and SCHNEIDER, K. H.: 'A critical comparison of artificial pollution test methods for HV insulators', *Electra*, 1979, **64**

121 LAMBETH, P. J., LOOMS, J. S. T., SFORZINI, M., MALAGUTI, C., PORCHERON, Y., and CLAVERIE, P.: 'International research on polluted insulators'. CIGRE, 33-02, 1970

122 CLAVERIE, P., and PORCHERON, Y.: 'How to choose insulators for polluted areas', *IEEE Trans.*, 1973, **PAS-92**, pp. 1122–1131

123 Extract from specification of a Middle Eastern supply authority

124 MOORE, W. J.: 'Physical chemistry' (Longmans, London, 1965), p. 361

125 HAMPTON, B. F.: 'Flashover mechanism of polluted insulation', *Proc. IEE.*, 1964, **111**, pp. 985–998

126 KING, L. A.: 'The voltage gradient of the free-burning arc in air or nitrogen'. 5th Internat. Gas Discharge Conf., Munich, p. 871 (North Holland Pub. Co.) Vol. 1, 1961

127 OBENAUS, F.: 'Fremdschichtueberschlag und kriechweglaenge', Deutsche Elektrotechnik 1958, **4**, pp. 135–136

128 NEUMAERKER, G.: 'Verschmutzungszustand und kriechweg', *Mon. Deut. Akad. Wiss., Berlin*, 1959, **1**, pp. 352–359

129 ALSTON, L. L., and ZOLEDZIOWSKI, S.: 'Growth of discharges on polluted insulation', *Proc. IEE.*, 1963, **110**, pp. 1260–1266

130 RIZK, F. A. M.: 'Mathematical models for pollution flashover', *Electra*, 1981, **78**, pp. 71–103

131 BOYLETT, F. D. A., and MACLEAN, I. G.: 'The propagation of electric discharges across the surface of an electrolyte', *Proc. Roy. Soc.*, 1971, *324*, **A**, pp. 469–489

132 MARSH, J. D.: CERL Rep. 1973. Quoted in Reference 48

133 SWIFT, D. A.: 'Arresting arc propagation with narrow metal strips', *Proc. IEE*, 1980, **127A**, pp. 553–564.

134 MAIKOPAR, A. S.: 'The open low current arc', *Elektrichestvo*, 1965, **2**, pp. 22—25

135 HURLEY, J. J., and LIMBOURN, G. J.: 'Correlation of service performance of insulators with test results'. CIGRE, SC 33, 1969

136 HESKETH, S.: 'General criterion for prediction of pollution flashover', *Proc. IEE*, 1967, **114**, pp. 531—532

137 WILKINS, R., and AL-BAGHDADI, A. A. J.: 'Arc propagation along an electrolyte surface', *Proc. IEE*, 1971, **118**, pp. 1886–1892

138 WOODSON, H. H., and McELROY, A. J.: 'Insulators with contaminated surfaces', *IEEE Trans.*, 1970, **PAS-89**, p. 1858

139 JOLLY, D.C., CHENG, T.C. and OTTEN, D.M.: 'Dynamic theory of discharge growth over contaminated insulator surfaces.', IEEE. Paper CP 74068-3, 1974.

140 RAHAL, A.M. and HURAUX, C.: 'Flashover mechanism of high voltage insulators.', *IEEE. Trans.*, 1979, **PAS-98**, 6, pp. 2223–2231.

141 SWIFT, D.A.: 'Water-triggered breakdown of an atmospheric air gap.', 1983. 4th ISH, Athens. 44-07.

142 ELY, C.H.A., LAMBETH, P.J. and LOOMS, J.S.T.: 'The booster shed.', *IEEE. Trans.*, 1978, **PAS-97**, 6, pp. 2187–2197.

143 FORREST, J.S., LAMBETH, P.J. and OAKESHOTT, D.F.: 'The performance of high voltage insulators in polluted atmospheres.', *Proc. IEE.*, 1960. *107A*, pp. 172–187.

144 VERMA, M.P.: 'Insulation performance of dc apparatus housing under pollution.', 1983. ETZ Archiv Bd.5, H.9., pp. 269–275.

145 KNUDSEN, N. and HERMANSSON, L.: 'Long and short term tests on polluted insulators.', CIGRE, 25-02, 1968.

146 KAWAMURA, T., NAGAI, K., SETA, T. and NAITO, K.: 'DC pollution performance of insulators.', CIGRE, 33-10, 1984.

147 ELY, C.H.A. and ROBERTS, W.J.: 'Switching impulse flashover of air gaps and insulators in an artificially polluted atmosphere.', *Proc. IEE.*, 1968, *115*, 11, pp. 1667–1671.

148 SETA, T.: 'Surge flashover tests on contaminated insulator.', 1961. CRIEPI, Japan, Rep. E-6001.

149 LUSHNICOFF, N.L. and PARNELL, T.M.: 'The effects of pollution and surface discharges on the impulse strength of line insulation.', IEEE. 71 TP 143-PWR, 1971.

150 YAMAZAKI, K., MITA, N., TOMIYAMA, J., and MIYOSHI, Y.: 'Countermeasures against salt pollution on insulators used for EHV transmission system located near and along sea coast'. CIGRE, 412, 1958

151 FORREST, J. S.: 'The characterstics and performance in service of high voltage porcelain insulators', *J. IEE*, 1942, **89**, Pt. II, pp. 60–80

152 ERLER, F.: 'Zum kriechueberschlag dicker isolatoren bei wechselspannung', *Elektrie*, 1969, **3**, pp. 100–102

153 VERMA, M. P., NIKLASCH, H., HEISE, W., LIPKEN, H., LUXA, G. F., and SCHREIBER, H.: 'The criterion for pollution flashover and its application'. CIGRE, 33-09, 1978

154 VERMA, M. P., and NIKLASCH, H.: 'Tests on insulator fitted with Raychem creepage extenders', 1984. Private communication

155 ZHANG RENYU, ZHU DEHENG, and ZHANG BINGYUAN.: 'The effect of arc bridging between sheds'. 4th ISH, ATHENS. 46.05, 1983

156 LOOMS, J.S.T.: 'Choosing insulators for arid regions', *Middle East Elect.*, June/July, 1984, pp. 38–41

157 LAMBETH, P. J.: 'Flashover of polluted HV insulators in a variety of wetting conditions' *in* 'Breakdown and discharge phenomena'. IEE Colloquium, London, 1975

158 FORREST, J. S.: 'The electrical characteristics of 132 kV line insulators under various weather conditions', *J. IEE*, 1936, *79*, pp. 401–413

159 REVEREY, G, and STOLTE, E.: 'Tests on insulators under natural conditions of contamination'. CIGRE 210, 1960

160 HEBERT, P. Y.: 'Une experience de cinq ans sur le comportement des isolateurs', *Rev, Gen. Elect.*, 1969, **78**, pp. 411–416

161 TEMPELAAR, H.: 'Results of natural pollution tests in Velsen'. CIGRE SC33-04, 13 IWD, 1969

162 REVEREY, G., and VERMA, M.P.: 'Natural pollution test station Friesenheimer Insel'. CIGRE SC33-04, 17 IWD, 1970

163 SPARROW, L. J.: 'Pollution performance of surge arrestors'. IEE Conf. on Lightning and power systems, London, 1984, pp. 96–101

164 ALLEN, L. J., and LE BAUT, P. Y.: 'Raccordement optique en haute tension'. RGE No. 6, 'Fibres optiques', 1986, pp. 18–23

165 BICC METALS Ltd.: 'Fibral conductors'. Publ. No. 980, 1983

166 ELY, C. H. A., and LAMBETH, P. J.: 'Further work on pollution characteristics of 400 kV substation insulators', *Proc. IEE*, 1971, **118**

167 RIZK, F. A. M., and BOURDAGES, M.: 'Influence of AC source parameters on flashover characteristics of polluted insulators'. IEEE 84 SM 709-2, 1984

168 VON CRON, H.: 'Testing insulators with reproducible foreign layers on their surface'. CIGRE 203, 1956

169 AKAZAKI, M., and TSUNEYASU, I.: 'Flashover on electrolyte surface at impulse voltage application', *Elec. Eng. Japan.*, 1968, **88**, pp. 46–54

170 TOMINAGA, A.: 'Characteristics of power frequency flashover on contaminated surfaces in fog', 1968, *ibid.* pp. 53–59

171 LEVSHUNOV, R. T., and NOVIKOV, A. A.: 'Pollution flashover of HV insulators', *Elektrichestvo*, 1959, **1**, pp. 84–88

172 SMAKALOVA, J.: 'Elektricke vlastnostil Francouzskeho izolatoru Spirelec SSb 10054', *EGU. Bulls.*, 1968, **5–6**, pp. 68–70

173 MEIER, H.: 'Examen du comportement d'isolateurs HT', *Rev. Gen Elect.*, 1969, **78**, pp. 417–428

174 HEISE, W., and LUXA, G. F.: 'Die bemessung der aeusseren isoliertrecken von geraeten fuer hoehe betriebsspannungen', *Electrotech. Zeit. A*, 1970, **81**, pp. 218–229

175 LAMBETH, P. J.: 'Effect of pollution on high voltage outdoor insulators', *Proc. IEE Reviews*, 1971, **118 (9R)**

176 MARSH, J. D.: CERL Internal Report, 1978

177 NASSER, E.: 'Verhalten von isolatoren bei underschiedlich verteilter fremdschicht', *Elektrotech. Z A*, 1963, **84**, pp. 353–357

178 BARBER, P. B., LOOMS, J. S. T., SWIFT, D. A., and TOZER, B. A.: 'Plasma homopolar'. *Brit. J. Appl. Phys.*, 1964, **15**, p. 665

179 CERL Internal Report, 1978

180 REVEREY, G.: 'Hochspannungs isolatoren unter fremschichteinfluss', *Elektrizitaetswirtschaft*, 1959, **58**, pp. 90–96

181 VDE 0448: 1968 (draft)

182 ISSEL, G., and BOEHME, H.: 'Zur pruefung von isolatoren mit zellulosefremdschichten', *Elektrie*, 1966, **1**, pp. 13–16

183 KOHOUTOVA, D.: 'Selection of insulators used in the CSSR for polluted areas'. CIGRE SC33 Colloqium, A-1, 1977

184 FUJITAKA, S., KAWAMURA, T., TSURUMI, S., KONDO, H., SETA, T., and YAMAMOTO, M.: 'Japanese method of artificial pollution test on insulators', *IEEE Trans.*, 1968, **PAS-87**, pp. 729–735

185 TAGAKI, T., HIROSE, Y., and HATTORI, H.: 'Flashover characteristics of large insulators'. CIGRE 25-03, 1968

186 KAWAI, M., and MILONE, D. M.: 'Flashover tests at project UHV on salt contaminated insulators', IEEE Trans., 1970, **PAS-89**, pp. 755–761

187 MACCHIAROLI, B. and TURNER, F. J.: 'A new contamination test method', IEEE Trans., 1969, **PAS-88**, pp. 1400–1408

188 MACCHIAROLI, B., and TURNER, F. J.: 'Study of some variables affecting contamination testing', *IEEE Trans.*, 1970, **PAS-89**, pp. 761–770

189 HEISE, W., and KOETHE, H. K.: 'Ueber das isoliervermoegen von ungleichmaessig verschmutzten hochspannungsisolatoren', *Elektrotech. Z A*, 1966, **87**, pp. 777–782

190 HEISE, W., LUXA, G. F., REVEREY, G., and VERMA, M. P.: 'Assessment of the solid layer artificial pollution test'. CIGRE 33-09, 1974

191 LAMBETH, P. J., KUIPERS, J. B., and JUMAH, A.: 'High voltage insulation for Bahrein and the problem of desert pollution'. Bahrein Soc. Eng. Seminar on Engineering in the Gulf, 1975

192 ALEKSANDROV, G. N., and KIZEVETTER, V. E.: 'Electric strength investigations of long strings of insulators', *Elektrotekhnika*, 1965, **36**, pp. 55–58

193 VERMA, M. P.: 'Insulator designs for 1200 kV lines', *IEEE Trans.*, 1981, **EL-16**, p. 3

194 LAMBETH, P. J.: 'Praca izolatorow w warunkach zabrudzeniowych', *Przeglad Elektrotech.*, 1969, **45**, pp. 28–31

195 LOOMS, J. S. T.: 'Live working on HV lines', *IEE Proc.*, 1981, **128A**, pp. 89–106

196 BRZUSKA, L.: 'Ueberschlagfestigkeit ungestuerter und gestuerter durchfuehrungen unter fremschichteinfluss', *AEG Mitt.*, 1965, **55**, pp. 22–24

197 PARGAMIN, L., HUC, J., and TARTIER, S.: 'Considerations on the choice of the insulators for HVDC overhead lines'. CIGRE 33–11, 1984

198 KJØLBY, A.: Correspondence between CERL and Norden Insulator Co, Copenhagen, 1970
199 Letter to author from Israel Electricity Corp., 1986
200 STEYER, F.: 'Ueber die aussenform von stutzern und langstabisolatoren', *Electrotech. Z.*, 1944, **65**, pp. 238–241
201 POHL, Z.: 'Znaczienie i dobor parametrow konstrukcjnych kloszy izolatorow', *Prezeglad Elektrotech.*, **43**, pp. 355–361
202 ROBINSON, W. G.: 'Developments in porcelain insulators', *Elect. Rev.*, 6 Oct, 1972, pp. 435–438
203 LOOMS, J. S. T.: 'Insulator problems are nothing new', *Asian Elect.*, Sept. 1984, pp. 49–51
204 GROENEWALD, H.: 'CAD of HV insulators'. 4th ISH, Athens, 11.01, 1983
205 SWIFT, D. A.: Private communication, CERL, Leatherhead, UK, 1981
206 ROBERTS, W. J., and LAMBETH, P. J.: 'The vulnerability of glass insulators to damage and failure', CERL L/N 74/76, 1976
207 KARADY, G., VINET, R., and SOUCHEREAU, N.: 'New test methods for synthetic insulators'. CIGRE 22-15, 1976
208 IEC TC 36: 'Tests on composite insulators for AC overhead lines'. (Draft) 6.3.2, 1986
209 SWIFT, D. A.: Communication to Tripartite Insulator Group, CEGB, EdF, ENEL, 1980
210 VDEW e.V.: 'Stoerungs- und schadensstatistik' 1978. (Verlags- und wirtschaftsgesellschaft der elektrizitaetswerke MBH, VWEW, 6 Frankfurt/Main, 70, 1978)
211 SPORN, P., CAHEN, F., and MAGNIEN, M.: 'Progress report of SC 9 on EHV AC transmission'. CIGRE Report 420, 1964
212 NGK INSULATORS LTD.: 'Hot line insulator washing equipment'. Technical Note TN 70048. Mizuho, Nagoya, Japan, 1970
213 LAMBETH, P. J., LOOMS, J. S. T., STALEWSKI, A. C., and TODD, W. G.: 'Surface coatings for HV insulators in polluted areas', *Proc. IEE.*, 1966, **113**, pp. 861–869
214 DOW CORNING INT.: 'Protect your HV insulators with silicone compounds'. Form No. 22-140-01, 1973
215 SETA, T.: 'Combating the pollution problem by applying grease-like water repellent substance', *Electrotech. J. Japan*, 1962, **7**, pp. 73–77
216 HALL, J., and ORBECK, T.: 'RTV coating for porcelain insulators'. IEEE 82 SM 363 – 0, 1982
217 ATKINS, A. D., LOOMS, J. S. T., ROBLES, E., and ROWE, J.: 'Improved mastic sealant' (Raychem Ltd., Swindon, 1985)
218 VERMA, M. P., and NIKLASCH, H.: 'Further tests on creepage extenders'. Tests 1/60 to 1/64. Raychem Ltd., Swindon, 1986
219 FORREST, J. S., and MARSHALL, C. W.: 'Improvement relating to HV insulators'. Brit. Patent 527 357, 1940
220 VOSE, W.: 'Semiconducting glaze'. Brit. Patent 577 748, 1946
221 GILLAM, G. H.: 'Stabilized insulators', *Elect. Times*, 1947, **112**, pp. 289–293
222 LEVSHUNOV, R. T.: 'Semiconductor glazed insulators', *Elektrichestvo*, 1969, **10**, 82–84
223 FUJIMURA, T., NAITO, K., and IRIE, T.: 'Performance of semiconducting glaze insulators under adverse conditions', *IEEE Trans.*, 1978, **PAS-97**, pp. 763–771
224 BINNS, D. B.: 'Use of valency controlled semiconducting oxides and development of tin-oxide glaze', *Trans. Brit. Ceram. Soc.*, 1974, **73**, p. 7
225 BINNS, D. B.: 'Improvements in and relating to glazes for ceramic articles', Brit. Patent, 982 600, 1965
226 POWELL, D. G.: 'Developments in glazes'. Private communication, Joslyn Mfg, Co., Lima, NY, USA, 1979
227 DOULTON & Co.: 'Improvements relating to electrical insulators'. Brit. Patent 1 098 958, 1966
228 NIGOL, O., REICHMAN, J., and ROSENBLATT, G.: 'Development of new semiconductive glaze insulators', *IEEE Trans.*, 1973, **T-73**, pp. 420–427

229 GOLDRING, A. G., HARTSHORN, P. R., RICKETTS, C. E., and ROBINSON, W.: 'Insulation for high-voltage AC. railway electrification in Great Britain', *Proc. IEE*, 1969, **116**, pp. 1377–1386

230 WHEELER, J. C. G.: 'Testing of solid core insulators for use on BR 25 kV electrification', *Proc. IEE*, 1983, **130B**, pp. 278–283

231 WHEELER, J. C. G.: 'Anti-vandal shed protectors for solid core porcelain insulators on British Rail'. DMMA Lancaster, 1984, pp. 310–313

232 MOREY, G. W.: 'The properties of glass' (Reinhold Publishing Corp., NY, 1954) 2nd edn., p. 127

233 KIRK OTHMER.: 'Encyclopedia Chemical Technology'. 'Paint and varnish removers' (Wiley Interscience, 1981) 3rd edn., pp. 762–768

234 LITTLETON, J. T., and MOREY, G. W.: 'The electrical properties of glass' (John Wiley, NY, 1933)

235 KIMOTO, I., FUJIMURA, T., and NAITO, K.: 'Performance of insulators for DC. transmission line under polluted conditions'. IEEE T72 556–9, 1972

236 EPRI: 'Transmission line reference book: HVDC to + 600 kV'. 1978

237 KAWAMURA, T., NAGAI, K., SETA, T., and NAITO, K.: 'DC pollution performance of insulators'. CIGRE 33-10, 1984

238 PARGAMIN, L., HUC, J., and TARTIER, S.: 'Consideration on the choice of the insulators for HVDC overhead lines'. CIGRE 33-11, 1984

239 BICC LTD.: 'Improvements relating to electric insulators'. Brit. Patent. 641 040, 1960

240 HIRSCH, F., von RHEINHABEN, H., and SORMS, R.: 'Flashovers of insulators under natural pollution and HVDC', *IEEE Trans.*, 1975, **PAS-94**, pp. 45–50

241 HIRSCH, F.: 'TDL epoxy insulator under DC'. Private communication, 1980

242 BAUER, E., KAERNER, H., MUELLER, K. H., and VERMA, M. P.: 'Service experience with the German longrod insulator with silicone rubber sheds since 1967'. CIGRE 22-11, 1980

243 BAKER, W. P., and VOAS, B.: 'Stress modification by fitments to polymeric insulator'. 1986. In course of publication

244 BSI.: 'Code of practice for abatement of radio interference from overhead power lines'. BS 5602, 7.3.3, 1978

245 DEMBINSKI, E. M.: 'RI properties of strings containing both standard-glazed and resistive-glazed units'. CERL internal communication, 1980

246 CORTINA, R., and SFORZINI, M.: 'Assessment of the RIV performance of insulators for high voltage application'. CIGRE WG. 36-01, 1982

247 CIGRE Study Cttee. 36, WG 36-01.: 'Comparison of radio noise prediction methods with CIGRE/IEEE survey results', *Electra*, 1972, (22) p. 175

248 CORTINA, R., de MICHAELIS, F., SFORZINI, M., and ZAFFANELLA, L.: 'Laboratory methods for measuring RI from insulator strings'. CISPR, WG3, Prague, 1966

249 MAXSTADT, F. W.: 'Surface breakdown', *Elect. Eng.*, 1934, **53**, p. 1062

250 BERNARDELLI, P. D., CORTINA, R., and SFORZINI, M.: 'RI performance of insulators in different ambient conditions', *IEEE Trans.*, 1973, **PAS—92**, p. 1

251 DE MICHAELIS, F., and ROSA, F.: 'Results of RIV measurements on cleaned and polluted insulators for different values of relative humidity'. CIGRE WG 36-01, 1982

252 BARBER, P. B.: CERL, private communication, 1980

253 EDF, SERVICE AEE.: 'Essais acoustiques de chaines d'isolateurs en soufflerie'. E22 KO3, 1978

254 ARBEY, H., DELCAMBRE, J., MICHAUD, R., MOREAU, M., PARGAMIN, L., and PAROT, J. M.: 'Les bruits éoliens des lignes électriques'. CIGRE Stockholm Symposium 232-05, 1981

255 LOOMS, J. S. T.: 'Whistle stop for insulators', *Elect. Rev.*, 1986, **219**, p. 31

256 LANGE, F. F.: 'Transformation toughening in the AL_2O_3/ZrO_2 composite system'. Rockwell Technical Report, 7, 1979

257 PAUL, A.: 'Glass-ceramics' *in* 'Chemistry of glasses' (Chapman & Hall, 1982) p. 41

258 BACHE, H. H.: 'Cement bound materials with extremely high strength and durability'. Aalborg Portland, POB 165, DK 9100, 1980

259 DOUBLE, D. D.: 'Cement – a respectable material?', *Nature*, 1981, **289**, pp. 348–349

260 MIYACHI, I., MORIYA, T., *et al.*: 'Seismic analysis and test on transformer bushings'. CIGRE 12-06, 1984

261 VERMA, M. P., and PETRUSCH, W.: 'Minimum requirements for AC test circuits giving uninfluenced results of tests on polluted insulators'. CIGRE Task Force 33.04.01 (Cairo Meeting), 1983

Appendix A

Glossary of insulator names

Function	Name	
	Europe	USA
Tie	Cap-and-pin disc String insulator	Suspension
Tie	Motor insulator	Short-rod
Tie	Longrod Langstab	Longrod
Tie	Polymeric insulator	Nonceramic Composite
Strut	Pin	Pin
Strut	Pedestal post Polyped	Cap-and-pin
Strut	Solid-core post Lapp-type post	Station post
Strut	Multiple-cone post Multicone post	Multicone
Housing	Bushing	Apparatus bushing
Housing	Housing	Special
Beam	Insulating crossarm	Composite post
Beam	Line post	Line post

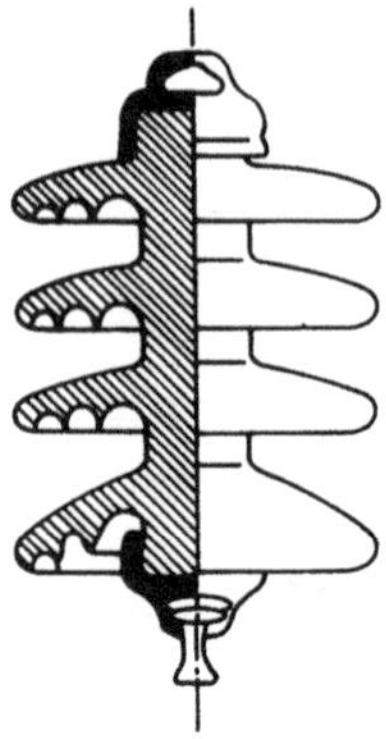

270 mm
Motor

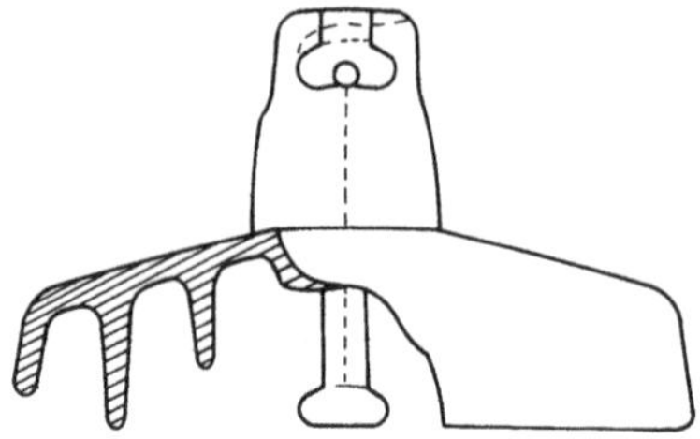

270 mm
disc

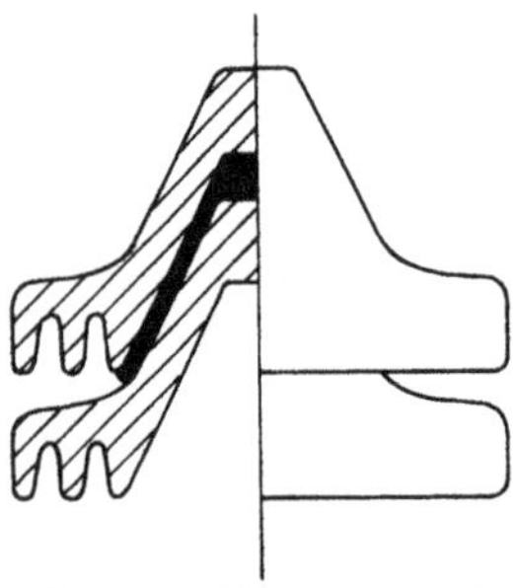

420 mm
multiple cone

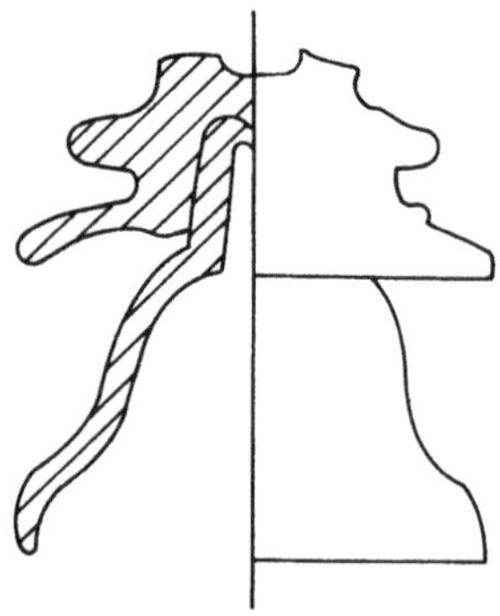

120mm
pin

Insulator nomenclature

The principal shapes of insulator are shown. Vertical-loaded strings of cap-and-pin discs are known, in England, as 'suspension sets' and when horizontal-loaded as 'tension sets'. For horizontal duty the shape is usually modified towards lesser convolution.

'Pedestal posts' are either assembled from single units of metal and porcelain, bolted together, or in the 'polyped' form, shown on page 261. Here sub-assemblies using continuous metal parts and embodying several units of metal and porcelain are bolted together.

The 'solid-core' post is of continuous porcelain with metal fittings, like the 'longrod' and 'Motor' but of much larger diameter. Its appearance is much the same as that of the 'multiple cone' which, as shown, comprises both porcelain and cement.

'Bushings' and 'housings' are shown in Figs. 14.3 and 14.6(b), respectively. 'Polymeric' insulators resemble longrods or line-posts but are always slimmer: see Figs. 9.1 and 9.3.

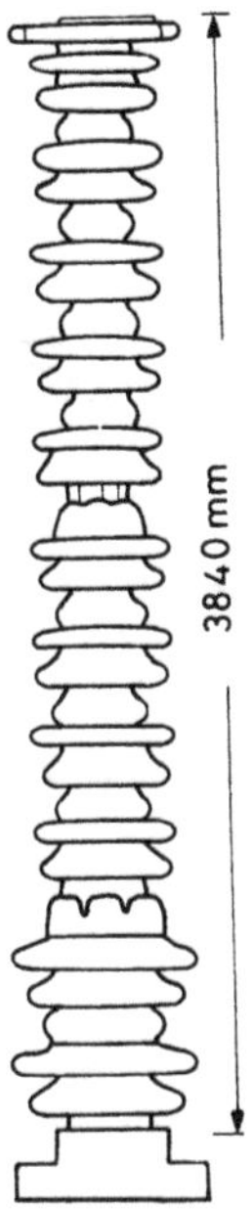

pedestal post

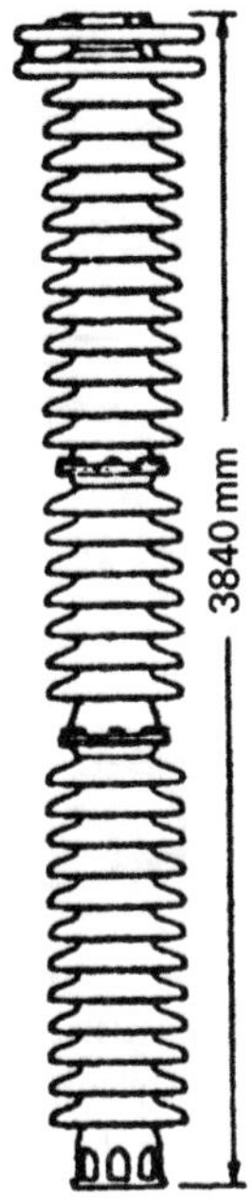

solid-core post

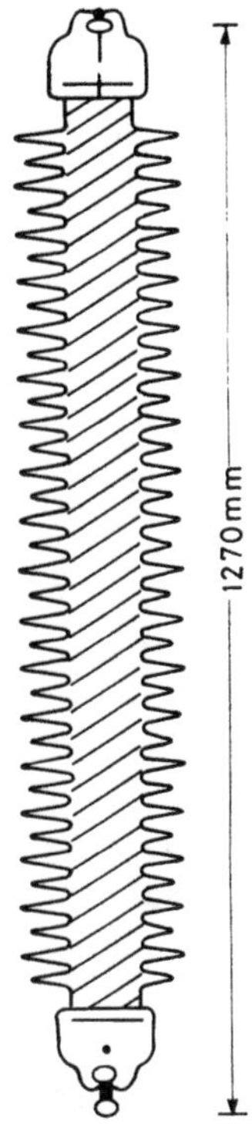

longrod

Appendix B

Testing of insulators

Tests in course of manufacture

For glass and porcelain, these include routine sampling of raw materials and physical tests on fired or moulded pieces, such as measurement of cross-breaking strength on small porcelain rods. Porcelain cap-and-pin bodies are usually tested routinely for electrical puncture strength and resistance to hydraulic bursting, while hollow shells are electrically puncture-tested only. Simple 'ringing' impact tests are made on porcelain components of multiple-cone posts, to detect cracks.

Polymeric insulators are not easy to test non-destructively although many types are proof-pulled, as are porcelain discs and longrods, both to reject weak pieces and to 'set' conical joints.

Standards

Some of the many Standards against which insulators of different types are tested are listed in Table B.1.

For glass and porcelain, the principal Standards are BS 137 (British), ANSI C 29 (American) and IEC 383 (International). A comparison of the latter two is given in Table B.2.

At the time of writing only a draft IEC Specification is available for polymeric insulators, the final Standard being not yet agreed. Similar action is in progress through an American working group on non-ceramic insulators.

Table B.1 *Standards for porcelain and glass insulators*

Body	Number	Year	Title
General			
ASTM	A-239-73	1973	Locating thinnest spot in Zn coating Preece test
	A-153-78	1978	Zn coating (hot dip)
BSI	923	1972	Guide on HV test techniques
CISPR	1	1975	Specification for RI measuring apparatus for frequency range 0·15 – 30 MHz
	7B	1975	Significance of a CISPR limit
NEMA	107	1976	Methods of measurement of RIV
IEC	60		HV test techniques
	60-1	1973	General definitions and test requirements
	60-2	1973	Test procedures
	60-3	1976	Measuring devices
	60-4	1977	Application guide for measuring devices
	437	1973	RI tests on HV insulators
	507	1975	Artificial pollution tests on HV insulators
	672		Specification for ceramic and glass insulating materials
	672-1	1980	Definitions and classification
	672-2	1980	Methods of testing
ANSI	C29-1	1982	Test methods for electrical power insulators
	C29-2	1983	Wet process porcelain and toughened-glass insulators
	C29-7	1977	HV-line post (insulator) types
	C29-9	1971	Insulators, apparatus, post type
	C68-1	1968	Techniques for dielectric tests
	C76-1	1976	Outdoor apparatus bushings
	C77-1	1943	American standard for wet tests
Disc insulators for strings			
IEC	120	1977	Dimensions of ball-and-socket couplings
	305	1978	Characteristics of string insulators of the cap-and-pin type
	372		Locking devices for ball-and-socket couplings
	372-1	1977	Dimensions and general rules
	372-2	1976	Tests
	383	1983	Tests on ceramic/glass insulators for voltage greater than 100 V
	438	1973	Tests and dimensions for HV DC insulators
	471	1977	Dimensions of clevis-and-tongue couplings
	506	1975	Switching impulse tests
	575	1977	Thermal mechanical endurance test

	591	1978	Sampling rules and acceptance criteria
BSI	137(1)	1970	Insulators of ceramic or glass for voltages greater than 1000 V: tests
	137(2)	1973	Same title: requirements
	3288	1973	Insulators and conductor fittings for overhead power lines: performance and general requirements: dimensions

Table B.2 *Principal test parameters for insulators as contained in ANSI C29-2 or IEC 383*

Type tests	ANSI	IEC
1 Dimensions	*	*
2 Power-frequency flashover voltage, dry	*	*
3 Same, wet (vertical or horizontal)	*	*
4 Lightning impulse flashover voltage, dry (both polarities)	*	*
5 Lightning impulse-withstand voltage, dry (both polarities)	*	*
6 Switching-surge voltage withstand, wet	No	*
7 Radio-influence voltage	*	No
8 Visual corona-extinction voltage	No	*
9 Time versus mechanical load	*	No
10 Thermal shock	*	*
11 Resistance to impact	*	No
12 Residual mechanical strength	*	No
13 Cotter key	*	*
Quality conformity tests		
14 Visual/dimensional	*	*
15 Porosity of porcelain fragment	*	No
16 Quality of galvanising	*	*
17 Combined mechanical & electrical	*	No
18 Electrical puncture	*	*
Routine tests		
19 Cold–hot thermal shock	*	*
20 Hot–cold thermal shock	*	*
21 Tension proof	*	*
22 Flashover/puncture	*	*

Note: There are minor differences between ANSI and IEC on impulse waveforms (electrical) and numbers of cycles (mechanical and thermal shock)

Appendix C

Selective bibliography on live washing of insulators

1 BUFFET, *et al.*: 'Playing fire-hoses on live conductors', *RGE*, 1934, **36**, pp. 305–316
An early analysis of leakage along water-jets of different conductivities. Table shown of safe distances.

2 HOLMSTROM, G.: 'Spraying of live line with sea water', *Brandkaarstidskrift*, 1937, **19**, pp. 189–196
Safe spraying distances using sea-water. (If Baltic water, much lower conductivities than from normally encountered salinities.)

3 ESTORFF, W., *et al.*: 'Water cleaning of live HV equipment', *ETZ*, 1940, **61**, pp. 817–827
Use of high pressures to break jets up into sprays. Details of nozzles.

4 EARDLEY, M.V.: 'Insulators washed on energised 132 kV line', *Elect. World*, 1944, **121**, pp. 78–80

5 BREMNER, W. C., *et al.*: 'Live 220 kV line insulators washed to stop leakage', *Elect. World*, 1944, **122**, pp. 90–92
Use of high-pressure jet washing to combat, not flashover, but leakage currents causing pole fires.

6 HILL, G. L.: 'Test developments with hot-line washing', *AIEE Trans.*, 1947, **66**, pp. 1203–1216
Recommends maintaining solid jet to impact point on insulator. Opposed to Reference 3.

7 POGGENDORF, A.: 'Cleaning live HV equipment with built-in sprinkler systems', *Elek. Wirts.*, 1951, **50**, pp. 31–36
Experience over 10 years with built-in sprinklers.

8 DANISH POWER ASSOCIATION COMMITTEE: 'Water spray appliances', *Elektroteknikeren*, 1952, **48**, pp. 101–109
Good account of involuntary live washing by agricultural spray plants. The hazards for 110 kV lines.

9 ALLEN, H. L.: 'Insulator washing by remote control', *Elect. Light & Power*, 1952, **30**, pp. 119–122
Hydraulic telescopic crane for pole-top washing at 66 kV.

10 KIAY, H.: 'The trajectory of water particles during spray tests on insulators', *ASE Bull.*, 1954, **45**, pp. 693–696
Demonstration of large effects of electric fields around insulators on droplet trajectories.

11 DEJUHASZ, K. J.: 'Spray literature abstracts', 1959. (Amer. Soc. Mech. Eng., NY, 1959)
1300 entries on many aspects of sprays.

12 BAATZ, H.: 'Spraying of HV lines with fire hoses', *ETZ B*, 1960, **12**, pp. 333–337
Jet must be atomised before striking insulator, for 110 kV.

13 FEDOTOV, M. P.: 'Cleaning live insulators by intermittent water jets', *Elek. Stanstii*, 1962, **6**, pp. 66–69
Recognises advantages of 'slugs' of water rather than solid jets or sprays.

14 LANE, J.: 'Live line insulator washing successful on EHV', *Elect. World.*, 1963, **160**, p. 104
Early success on 345 kV line insulators.
15 See Reference 106: LAST, F. H. *et al.*, 1966
16 STALEWSKI, A. C., and DAWKINS, R. B.: 'Remotely controlled live washing', *Elect. Rev.*, 1967, **180**, pp. 708–711
17 NGK INSULATORS LTD.: 'Hot line washing equipment'. Technical Note 70048. Mizuho, Nagoya, Japan, 1970
Detailed accounts of spray and artificial rain methods.
18 KNUCKEY, A.: 'Live line insulator washing', *Middle East Elect.*, March 1987, pp. 20–21
Fixed ring sprays for overhead lines up to 220 kV, especially in desert areas, using Mather Platt 'plug-in' system with mobile tankers.

This bibliography is based on work by D. C. Levitt, CEGB Information Services, London, England

Subject Index

Name Index

www.ingramcontent.com/pod-product-compliance
Lightning Source LLC
LaVergne TN
LVHW011750200926
848484LV00003B/664